Beginning Algebra I

Michael Gaul, Barbara Goldner, Edgar Jasso, Deanna Li,
Pam Lippert, Eileen Murphy, Sam Wilson

3rd Edition
Fall 2018

Contents

1 Linear Equations 9

1.1 Order of Operations — 9
- Exponents 11
- Fraction Bar 13
- Absolute Value 16

1.2 Algebraic Expressions — 19
- Evaluating Algebraic Expressions 20
- Simplifying Algebraic Expressions 21
- Distributive Property 23
- Negative Sign in Front of a Parenthesis 24
- Commutative and Associative Properties 26

1.3 Property of Equality — 31
- Addition and Subtraction Property of Equality 32
- Multiplication and Division Property of Equality ... 35

1.4 Linear Equations - General Case — 41
- Variable on One Side of the Equation 41
- Variable on Each Side of the Equal Sign 42
- Equations with Fractions 47

Mid-Chapter 1 Check-Up — 53

1.5 Linear Inequalities — 54
- Solving Linear Inequalities 57

	1.6	**Absolute Value Equations**	64
	1.7	**Formulas**	70
		Formulas with Parenthesis	73
		Fractions in Formulas	75
	1.8	**Applications of Linear Equations & Inequalities**	81
		Problem Solving Strategies and Tools (PSST)	81
		Applications Involving Percents and Fractions	81
		Geometry	85
		Business Applications	88
		Inequalities	89

2 Graphing Linear Equations ... 95

	2.1	**Introduction to Graphing**	95
		Rectangular Coordinate System	95
		Graph of an Equation in Two Variables	99
	2.2	**Graphing Linear Equations in Two Variables**	103
		Vertical and Horizontal Lines	108
	2.3	**Intercepts**	116
		Graphing Lines Using Intercepts	118
		Mid-Chapter 2 Check-Up	126
	2.4	**Slope**	127
		Reading the Slope from a Graph	128
		Slope of Horizontal and Vertical Lines	132
		Slope Formula	135
		Using a Point and the Slope to Find Another Point	138
		Applications of the Slope	140
	2.5	**Slope-Intercept Form**	146
		Graphing Lines Using the Slope and the y-Intercept	149
		Finding the Equation of a Line from a Graph	153
		Equations of Horizontal and Vertical Lines	155
		Applications	157
	2.6	**Point-Slope Form**	164
		Finding the Equation of a Line given Two Points	166
		Applications	172
	2.7	**Parallel and Perpendicular Lines**	179
		Slopes of Parallel and Perpendicular Lines	179
		Horizontal and Vertical Lines	187
	2.8	**Linear Inequalities in Two Variables**	191
		Graphing Linear Inequalities in Two Variables	193
		Horizontal or Vertical Line Boundary	199

3 Systems of Linear Equations 207

3.1 Introduction to Systems of Linear Equations   **207**
      Systems of Two Linear Equations 208

3.2 Solving Systems of Linear Equations by Graphing   **213**

3.3 Substitution Method   **225**
      One Isolated Variable 225
      No Isolated Variable 228

3.4 Elimination Method   **236**

Mid-Chapter 3 Check-Up   **246**

3.5 Applications of Systems of Linear Equations   **247**
      Mixture Problems 247
      Distance, Rate and Time Problems 253

4 Exponents 259

4.1 Introduction to Exponents   **259**
      Positive Exponents 259

4.2 Properties of Exponents   **263**
      Product Rule 263
      Quotient Rule 264
      Power Rule of Exponents 266
      Power of a Product Rule 267
      Power of a Quotient Rule 268

4.3 Negative Exponents and Zero Power   **273**
      Zero Power 273
      Negative Exponents 276

4.4 Scientific Notation   **287**

5 Polynomials 295

5.1 Introduction to Polynomials   **295**
      Terms of a Polynomial 296
      Degree of a Polynomial 297

5.2 Addition and Subtraction of Polynomials   **300**
      Like Terms 300
      Adding and Subtracting Polynomials 302

5.3 Multiplication of Polynomials   **306**
      Multiplication by Monomials 306
      Multiplication of Binomials 308

5.4 Division of Polynomials   **315**
      Dividing by a Monomial 315
      Long Division 316

Answer Key 358

Preface

Mathematics instructors from North Seattle College (NSC) created this workbook to better serve students and to align the curriculum with NSC learning outcomes. This workbook assumes no prior knowledge of algebra, but assumes that students are familiar with the basic rules of arithmetic. This is the first part of a series of two workbooks: Beginning Algebra I and Beginning Algebra II. After mastering the topics included in this series students will be prepared for a course in Intermediate Algebra.
This workbook is organized as follows.

- **Chapters**

  ◇ **Sections** are the main instructional component for each chapter. This is where ideas are introduced.

  ◇ **Worked Examples** provide further explanation of a concept. It is recommended that students read and work through these examples carefully.

  ◇ **Class Examples** provide classroom examples for instructors as concepts are being developed.

  ◇ **You Try** are problems embedded in the workbook to help reinforce the concepts learned. It is recommended that students work through the problems in the order they appear, showing as much work as possible in a neat and organized fashion. Space for work is provided in the workbook. However, there may not be enough space so it is recommended that a notebook be used for all math work.

- **Section Exercises**

  ◇ Each section ends with a set of **exercises.** The only way to learn math is by practice. We suggest that students attempt the exercises on their own first before seeking help. There is no space to show work for the exercises. It is highly recommended that students keep an organized notebook for the class. Students should do the exercises in their notebook, showing all worked solutions in a neat and organized way, so they can refer to them easily.

◇ Solutions to the Exercises are at the end of the workbook. It is very important to verify that you have the correct answer before proceeding to the next problem.

Beginning Algebra I is licensed under a Creative Commons Attribution-NonCommercial-ShareAlike 4.0 International License.)
Based on a work by Tyler Wallace at http://wallace.ccfaculty.org/book/book.html

Student Writing Logo is based on a work by oksmith at openclipart.org and is licensed under a Creative Commons Universal Public Domain License (CC0 1.0)
https://creativecommons.org/publicdomain/zero/1.0/

1. Linear Equations

1.1 Order of Operations

Objective: To evaluate expressions using the order of operations

When simplifying expressions, it is important that we simplify them in the correct order as follows:

Parenthesis (Grouping Symbol)	**P**
Exponents	**E**
Multiply or **D**ivide (Left to Right)	**MD**
Add or **S**ubtract (Left to Right)	**AS**

Multiply and Divide are at the same level because they are the same operation (division is multiplication by reciprocal). This means that they must be done from left to right. So for some problems, we will divide first, while for others, we will multiply first. The same is true for adding and subtracting (since subtracting is adding the opposite).

Often, students use the word PEMDAS to remember the order of operations, as the first letter of each operation creates the word PEMDAS, as shown above. If we think about PEMDAS as a vertical word written as above we are more likely to remember that multiplication and division are done left to right (same as addition and subtraction).

Another way to remember the order of operations is to think of a phrase such as *"Please excuse my Dear Aunt Sally"* where each word starts with the same letter as the mathematical operation in the correct order.

Example 1 Simplify $4-7+5$

Solution.

$$4-7+5 = -3+5$$ 	Subtract 4 and 7 first
$$= 2$$ 	Our Solution

Example 2 Simplify $8+2\cdot 3$

Solution.

$$8+2\cdot 3 = 8+6$$ 	Multiply 2 and 3 first
$$= 14$$ 	Our Solution

Example 3 Simplify $30\div 3(-2)$

Solution.

$$30\div 3(-2) = \underbrace{30\div 3}(-2)$$ 	Divide first
$$= \underbrace{10(-2)}$$ 	Multiply
$$= -20$$ 	Our Solution

In the example above, if we had multiplied 3 and -2 first, the answer would have been -5, which is incorrect.

Exercise 1 Class Example
Simplify the following.

a) $-7-5+9$

b) $17-3\cdot 5$

c) $\dfrac{2}{5}(1-4+3)$

d) $(3-9)\div(5+17)$

It can take several steps to complete a problem. The key to successfully complete an order of operations problem is to take the time to show your work clearly. Do one step at a time. This will reduce the chance of making a mistake.

1.1 Order of Operations

Exercise 2 You Try

Simplify the following.

a) $5 - 12 - 3$

b) $7 \div (8 - 9)$

c) $8 - 2(4 - 9)$

d) $24 \div 4 \cdot 3$

Exponents

The next example illustrates an important concept about an exponent and its base. When we see -8^2, the exponent 2 has as its base, 8, and not -8. Meaning, only the 8 is squared, not -8. So, $-8^2 = -(8 \cdot 8) = -64$. However, when a negative number is in parenthesis, such as $(-8)^2$, it means that the exponent 2 has as its base, -8. Therefore, $(-8)^2 = (-8) \cdot (-8) = 64$. As a result, $(-8)^2$ gives a positive solution, 64, whereas -8^2 gives a negative solution, -64.

Example 4 Simplify $-4^2 - (-5)^2$

Solution.

$$
\begin{aligned}
-4^2 - (-5)^2 &= -4 \cdot 4 - (-5) \cdot (-5) && \text{Rewrite } 4^2 \text{ as } 4 \cdot 4 \text{ and } (-5)^2 \text{ as } (-5) \cdot (-5) \\
&= -16 - 25 && \text{Subtract} \\
&= -41 && \text{Our Solution}
\end{aligned}
$$

Example 5 Simplify $2 + 3(9 - 4)^2$

Solution.

$$
\begin{aligned}
2 + 3(9 - 4)^2 &= 2 + 3\underbrace{(9 - 4)}^2 && \text{Perform operation inside parenthesis} \\
&= 2 + 3(5)^2 && \text{Apply exponents} \\
&= 2 + 3\underbrace{(25)} && \text{Multiply} \\
&= \underbrace{2 + 75} && \text{Add} \\
&= 77 && \text{Our Solution}
\end{aligned}
$$

In the above example, if we had added 2 and 3 first, we would have gotten the answer 125, which is incorrect.

Exercise 3 Class Example
Simplify the following.

a) 9^2

b) -9^2

c) $(-9)^2$

d) $5-3(2-4^2)$

e) $6-(2-4)^2$

f) $(4-7)^2-(8-5)^2$

Exercise 4 You Try
Simplify the following.

a) $(-3)^2$

b) $-(-3)^2$

c) $5+(3-4)^2$

d) $5^2-(-6)^2$

1.1 Order of Operations

Fraction Bar

A type of grouping symbol that can be used in addition to the parenthesis is the fraction bar. In this case, the entire numerator and the entire denominator must first be evaluated until there is a single numerical value in each, the numerator and denominator. We can then simplify the fraction.

Example 6 Simplify $\dfrac{4-10}{18 \div 2}$

Solution

$$\dfrac{4-10}{18 \div 2} = \dfrac{-6}{9} \qquad \text{Numerator: Subtract; Denominator: Divide}$$

$$= -\dfrac{2}{3} \qquad \text{Our Solution in simplified term}$$

Example 7 Simplify $\dfrac{2^4 - 3(-8)}{15 \div 5 - 1}$

Solution.

$$\dfrac{2^4 - 3(-8)}{15 \div 5 - 1} = \dfrac{\overbrace{2^4} - 3(-8)}{\underbrace{15 \div 5} - 1} \qquad \text{Numerator: Apply exponent; Denominator: Divide}$$

$$= \dfrac{16 - \overbrace{3(-8)}}{\underbrace{3 - 1}} \qquad \text{Numerator: Multiply; Denominator: Subtract}$$

$$= \dfrac{\overbrace{16 + 24}}{2} \qquad \text{Numerator: Add}$$

$$= \dfrac{40}{2} \qquad \text{Simplify}$$

$$= 20 \qquad \text{Our Solution}$$

Note. Use extreme care when a division involves zero. If zero is divided by a number other than zero, we get zero.

- Example: $0 \div 2 = 0$. To check, multiply the divisor, 2, by the quotient (answer), 0, to get the dividend, 0. Check: $2 \cdot 0 = 0$.

- Example: $\dfrac{0}{-5} = 0$. Check: $-5 \cdot 0 = 0$.

However, if a number, other than zero, is divided by 0, the result is undefined.

- Example: $2 \div 0$ is undefined. Check: there is no number that can be multiplied by the divisor, 0, to get the dividend, 2.

- Example: $-\dfrac{9}{0}$ is undefined. There is no number that can be multiplied by the divisor, 0, to get the dividend, -9.

Example 8 Simplify $\dfrac{3^2-(4+5)}{7+5}$

Solution.

$$\dfrac{3^2-(4+5)}{7+5} = \dfrac{\overbrace{3^2}-(4+5)}{\underbrace{7+5}}$$ Numerator: Apply exponent; Denominator: Add

$$= \dfrac{9-\overbrace{(4+5)}}{12}$$ Numerator: Perform operation inside parenthesis

$$= \dfrac{\overbrace{9-9}}{12}$$ Numerator: Subtract

$$= \dfrac{0}{12}$$ Simplify

$$= 0$$ Our Solution

Example 9 Simplify $\dfrac{3(-5)}{7+2-9}$

Solution.

$$\dfrac{3(-5)}{7+2-9} = \dfrac{\overbrace{3(-5)}}{\underbrace{7+2}-9}$$ Numerator: Multiply; Denominator: Add

$$= \dfrac{-15}{\underbrace{9-9}}$$ Denominator: Subtract

$$= \dfrac{-15}{0}$$ Cannot divide by 0

Undefined   Our Solution

1.1 Order of Operations

Exercise 5 Class Example
Simplify the following.

a) $\dfrac{7+11}{4\cdot 2}$

b) $\dfrac{10+2\cdot 3}{1-(6-7)}$

c) $\dfrac{(8-10)^2}{3+6-9}$

d) $\dfrac{-4^2+16}{3(1-4)}$

Exercise 6 You Try
Simplify the following.

a) $\dfrac{3-11}{36\div 9}$

b) $\dfrac{(7-9)^2}{3\cdot 6-18}$

c) $\dfrac{15-5\cdot 3}{8(4+2)}$

d) $\dfrac{4-2(3+2)}{5^2-3^3}$

Absolute Value

Another type of grouping symbol is the absolute value sign. To simplify, we have to evaluate the expression inside the absolute value first until we get a single numerical value. Then, take the absolute value of the numerical expression. . Remember that the absolute value of a number is the distance from that number to the number zero, for example:

- $|5| = 5$ because the distance from 0 to 5 is 5 units.

- $|-5| = 5$ because the distance from 0 to -5 is 5 units.

Example 10 Simplify $|2-12|$

Solution.

$$|2-12| = |-10| \qquad \text{Perform subtraction inside absolute value}$$
$$= 10 \qquad \text{Our Solution}$$

Example 11 Simplify $|3 \cdot 2 - 9|$

Solution.

$$|3 \cdot 2 - 9| = |6 - 9| \qquad \text{Perform multiplication inside absolute value}$$
$$= |-3| \qquad \text{Subtract 9 from 6}$$
$$= 3 \qquad \text{Our Solution}$$

Example 12 Simplify $3|4-8| + 2|5+1|$

Solution.

$$3|4-8| + 2|5+1| = 3\underbrace{|4-8|} + 2\underbrace{|5+1|} \qquad \text{Perform operation inside absolute value}$$
$$= 3\underbrace{|-4|} + 2\underbrace{|6|} \qquad \text{Evaluate the absolute value}$$
$$= \underbrace{3(4)} + \underbrace{2(6)} \qquad \text{Multiply}$$
$$= \underbrace{12 + 12} \qquad \text{Add}$$
$$= 24 \qquad \text{Our Solution}$$

1.1 Order of Operations

Exercise 7 Class Example
Simplify the following.

a) $-3|-2|$

b) $7-|-5|$

c) $|4|-(6-10)$

d) $|9-6|+|3+8|$

Exercise 8 You Try
Simplify the following.

a) $|-5|+|8|$

b) $|6 \cdot 2 + 3|$

c) $|9-4|-(12-19)$

d) $|3^2 - 4^2|$

1.1: Exercises

Simplify the following.

1. $-6 \cdot 4(-1)$
2. $\frac{1}{2} - \frac{2}{3} + \frac{3}{4}$
3. $(-6 \div 6)^2$
4. $7 - 4 \cdot \frac{4}{5}$
5. $9 - |2 - 10|$
6. $8 \div 4 \cdot 2^2$
7. $(3-7)^3$
8. $3 + (8) \div |4 - 2|$
9. $11 - (2-4)^3$
10. $6 - \frac{1}{2}(5-8)$
11. $5 + 3(2 - 6 \cdot 4)$
12. $5(-5 + 6) - 6^2$
13. $1 + 3|2 + 4|$
14. $\frac{-3-1}{-2-(-2)}$
15. $\frac{1}{2} \cdot \frac{7}{3} - \left(\frac{4}{3}\right)^2$
16. $5^2 - (-8)^2$
17. $\frac{-10-6}{(-2)^2} - 5$
18. $4 - 2|3^2 - 16|$
19. $\frac{-5^2 + (-5)^2}{2^4 - 4 \cdot 3}$
20. $-6^2 + |-3 - 3|$
21. $\frac{2 - |7 + 2^2|}{4 \cdot 2 + 5 \cdot 3}$
22. $[-9 - (2-5)] \div (-6)$
23. $[-1 - (-5)] |3 + 2|$
24. $\frac{(-2+1) - (-3)}{-9 \cdot 2 - 3(-6)}$

Rescue Roody!

25. Roody was asked to simplify $2 + 5(6 - 3)$. This is what he did, but his work was marked incorrect. Help Roody understand what he did wrong.

$$\begin{aligned} 2 + 5(6-3) &= 2 + 5(3) \\ &= 7(3) \\ &= 21 \end{aligned}$$

26. Roody was asked to simplify $8 \div 4(2-7)$, but his answer does not match the answer in the back of the book. Help Roody.

$$\begin{aligned} 8 \div 4(2-7) &= 8 \div 4 - 5 \\ &= 2 - 5 \\ &= -3 \end{aligned}$$

1.2 Algebraic Expressions

Objective: To evaluate expressions by substituting given values, and simplify algebraic expressions by combining like terms

A **mathematical expression** is a mathematical phrase. In this section we will learn 2 types of mathematical expressions: numerical expressions and algebraic expressions.

- A **numerical expression** is a type of mathematical expression involving numbers that are connected by some arithmetic operation. Examples are $3(5)$ or $-2(6-4^2)$.

- An **algebraic expression** is similar to a numerical expression, but it also contains variables. **Variables** are symbols, usually denoted by letters, that represent unknown values. Examples of algebraic expression are $3x, p-4.7, -\frac{2}{5}x+8y+25$.

A **term** is a real number or a real number times a variable to a positive integer power. If there are two or more terms in an expression, they are separated by an addition or subtraction sign. Consider the expression, $p-4.7$. This expression contains 2 terms, namely p and -4.7. The expression $-\frac{2}{5}x+8y+25$ contains three terms, namely, $-\frac{2}{5}x, 8y$, and 25.

The **coefficient** of a term is the number that is multiplied by the variable. For example, in the term $3x$, 3 is the coefficient of the variable, x.

Exercise 1 **Class Example**
Given the expression $4p + \frac{q}{7} + r$.

a) How many terms are in the expression?

b) Identify the coefficients of each term.

Exercise 2 **You Try**
Given the expression $5x + 0.2y - z$.

a) How many terms are in the expression?

b) Identify the coefficients of each term.

> **World View Note** The term "Algebra" comes from the Arabic word al-jabr which means "reunion." It was first used in Iraq in 830 AD by Mohammad ibn-Musa al-Khwarizmi.

Evaluating Algebraic Expressions

Learning how to evaluate an expression is important in algebra. It becomes very useful when we need to check answers later. To evaluate an expression, replace a given value for each variable. It is always a good idea to put the substituted number in parenthesis, especially if the number is negative. This is to preserve mathematical operations that are sometimes lost in a simple replacement. Sometimes, a parenthesis won't make a difference but it is a good habit to always use them to prevent problems later.

Example 1 Evaluate the expression $p(q+6)$ when $p = 3$ and $q = 5$.

Solution.

$$\begin{aligned} p(q+6) &= (3)((5)+6) & &\text{Replace p with 3 and q with 5} \\ &= (3)(\underbrace{(5)+6}) & &\text{Evaluate terms inside parenthesis} \\ &= \underbrace{(3)(11)} & &\text{Multiply} \\ &= 33 & &\text{Our Solution} \end{aligned}$$

Exercise 3 Class Example
a) Evaluate the expression $2p - q$ when $p = 10$ and $q = -6$.

b) Evaluate the expression $\frac{1}{2}bh$ when $b = 12$ and $h = \frac{1}{3}$

Exercise 4 You Try
a) Evaluate the expression $2a - b + 5$ when $a = 3$ and $b = -4$

b) Evaluate the expression $2m + \frac{3}{4}n$ when $m = -7$ and $n = 12$

c) Evaluate the expression $\frac{3-k}{hk}$ when $h = 2$ and $k = -3$

Simplifying Algebraic Expressions

We can simplify algebraic expressions by combining like terms.

What are like terms? **Like terms** are terms whose variables match exactly, including exponents. Examples of like terms are $3x$ and $-7x$, $4a^2$ and $8a^2$, or 3 and -5.

How does combining like terms work? Consider shopping for fruit with a friend at a local farmer's market. You purchase 3 apples. Your friend purchases 2 apples. How many total apples were purchased? 3 apples plus 2 apples give us 5 apples. Similarly, $3a$ plus $2a$ gives us $5a$, or mathematically, $3a + 2a = 5a$. To combine like terms, add (or subtract) the coefficients, and keep the variables the same.

What are unlike terms? Returning to the shopping analogy, suppose you want to purchase 3 apples but your friend wants to purchase 2 oranges. There is no way to describe apples and oranges as the same thing. They are different items. The same holds true with variables. We cannot combine $3a + 2b$ since the variables do not match. $3a$ and $2b$ are unlike terms.

Similarly, $x + x^2$ are unlike terms because the exponents of x do not match. Recall that raising a base to an exponent is another way of expressing repeated multiplication. For example,

$3^2 = 3 \cdot 3 = 9$ (which is different from 3)

$5^2 = 5 \cdot 5 = 25$ (which is different from 5)

$x^2 = x \cdot x$ (which is different from x)

Consequently, we cannot combine $x + x^2$. These are unlike terms.

Example 2 Simplify $3x + 6x$

Solution.

$$3x + 6x = \underbrace{3x + 6x}_{} \qquad \text{Add coefficients of x}$$
$$= 9x \qquad \text{Our Solution}$$

Example 3 Simplify $2y + y^2$

Solution.

The terms are unlike. They cannot be combined. This expression is in simplified form.

Example 4 Simplify $5x - 2y - 8x + 7y$

Solution.

$$5x - 2y - 8x + 7y = \underbrace{5x - 8x}_{} \underbrace{- 2y + 7y}_{} \qquad \text{Combine like terms}$$
$$= -3x + 5y \qquad \text{Our Solution}$$

Example 5 Simplify $8a - 3b + 7.1 - 2a + 2.5b - 3$

Solution.

$$8a - 3b + 7.1 - 2a + 2.5b - 3 = \underbrace{8a - 2a}_{} \underbrace{- 3b + 2.5b}_{} + \underbrace{7.1 - 3}_{} \qquad \text{Combine like terms}$$
$$= 6a - 0.5b + 4.1 \qquad \text{Our Solution}$$

Exercise 5 Class Example
Simplify the following.

a) $7m - 4p + 12m - 18p$

b) $4.5m + n - 3.2 + 7.8m - 0.2n - 3$

1.2 Algebraic Expressions

Exercise 6 You Try
Simplify the following.

a) $4a - 3b - 7a + 9b$

b) $4.1p + 3q - 6 - 2.7p + 2.6q - 4.3$

c) $2n^2 + 2n$

d) $\dfrac{1}{2}x + 3y - 4 - 5x + \dfrac{8}{3}$

Distributive Property

The distributive property is another tool used to simplify algebraic expressions.

Distributive Property:
$$a(b + c) = ab + ac$$

Example 6 Multiply $4\left(2x - \dfrac{9}{2}\right)$

Solution.

$$4\left(2x - \dfrac{9}{2}\right) = 4(2x) - 4\left(\dfrac{9}{2}\right) \qquad \text{Multiply each term by 4}$$
$$= 8x - 18 \qquad \text{Our Solution}$$

Example 7 Simplify $-7(5x - 6)$

Solution.

$$-7(5x - 6) = -7(5x) + (-7)(-6) \qquad \text{Multiply each term by } -7$$
$$= -35x + 42 \qquad \text{Our Solution}$$

Negative Sign in Front of a Parenthesis

When a negative sign is in front of the parenthesis, it is the same as having a (-1) in front of the parenthesis. This means that we should distribute a (-1) to each term inside the parenthesis.

Example 8 Simplify $-(4x-5y+6)$

Solution.

$$\begin{aligned}-(4x-5y+6) &= -1(4x-5y+6) &&\text{Same as } -1 \text{ in front of parenthesis}\\ &= -1(4x)+(-1)(-5y)+(-1)(6) &&\text{Multiply each term by } -1\\ &= -4x+5y-6 &&\text{Our Solution}\end{aligned}$$

Note. Having a negative in front of the parenthesis has the same effect as changing the sign of each term inside the parenthesis.

Alternate Solution

$$-(4x-5y+6) = -4x+5y-6 \quad \text{Change the sign of each term inside the parenthesis}$$

Exercise 7 Class Example
Simplify the following.

a) $-(a+2b-3c)$

b) $-(2x-5y+1)$

c) $6(3m+n+7)$

d) $\dfrac{2}{3}\left(6a+3b-\dfrac{9}{4}\right)$

1.2 Algebraic Expressions

Exercise 8 You Try
Simplify the following.

a) $-(4k+7)$

b) $-(3x+4y-2)$

c) $-3(2g-h+9k)$

d) $4(5k+\frac{1}{3}p-\frac{1}{2})$

The next examples require both applying the distributive property and combining like terms.

Example 9 Simplify $5+3(2x-4)$

Solution.

$$5+3(2x-4) = 5+6x-12 \qquad \text{Distribute 3}$$
$$= 6x-7 \qquad \text{Our Solution after combining like terms}$$

Example 10 Simplify $3x-(4x-5)$

Solution.

$$3x-(4x-5) = 3x-4x+5 \qquad \text{Change sign of each term inside parenthesis}$$
$$= -x+5 \qquad \text{Our Solution after combining like terms}$$

Example 11 Simplify $4(3x-8)-(2x-7)$

Solution.

$$4(3x-8)-(2x-7) = 4(3x-8)-1(2x-7) \qquad \text{Distribute 4 and } -1$$
$$= 12x-32-2x+7 \qquad \text{Combine like terms}$$
$$= 10x-25 \qquad \text{Our Solution}$$

Exercise 9 Class Example
Simplify the following.

a) $3(4y-1)-(2y-5)$

b) $10x-16-6(4x+3)$

Exercise 10 You Try
Simplify the following.

a) $5m-(2m-n)$

c) $(3g-5)-(7-8g)$

b) $2(3p-4)+5$

d) $\frac{2}{3}(9n+6)+\frac{1}{4}(12-8n)$

Commutative and Associative Properties

There are two properties that help us simplify expressions.

> The **Commutative Property of Addition and Multiplication** states that the order of the numbers can be changed without changing the result. That is
>
> $$a+b=b+a \text{ and } ab=ba$$

1.2 Algebraic Expressions

> The **Associative Property of Addition and Multiplication** states that numbers can be grouped in any order without changing the result. That is,
> $$a+(b+c)=(a+b)+c \text{ and } a(bc)=(ab)c$$

Example 12 Simplify $(3x-1)4$

Solution.

$$(3x-1)4 = 4(3x-1) \quad \text{Reorder using commutative property}$$
$$= 12x-4 \quad \text{Our Solution after distributing 4}$$

Example 13 Simplify $4\left(\frac{3}{2}(p-7)\right)$

Solution.

$$4\left(\frac{3}{2}(p-7)\right) = \left(4 \cdot \frac{3}{2}\right)(p-7) \quad \text{Regroup using associative property}$$
$$= \left(4 \cdot \frac{3}{2}\right)(p-7) \quad \text{Multiply 4 and } \frac{3}{2}$$
$$= 6(p-7) \quad \text{Distribute 6}$$
$$= 6p-42 \quad \text{Our Solution}$$

Exercise 11 Class Example
Simplify the following.

a) $9\left(\dfrac{x+1}{3}\right)$

b) $-\dfrac{5}{3}(6(y+4))$

Exercise 12 You Try
Simplify the following.

a) $\dfrac{3}{4}(-8(2x+5))$

c) $-12\left(\dfrac{2k+3}{4}\right)$

b) $4\left(\dfrac{5}{8}(v-3)\right)$

d) $-5\left(\dfrac{3k-7}{5}\right)$

1.2: Exercises

Evaluate each expression using the values given.

1. $x(y-6)$; use $x=4$, $y=-1$
2. $2a-3b$; use $a=5$, $b=-2$
3. $(p-q)^2$; use $p=3$, $q=6$
4. $5(3a+6c)$; use $a=-3$, $c=\dfrac{1}{2}$
5. $|p+q|-(m+1)$; use $m=1$, $p=3$, $q=4$
6. y^2+y-z; use $y=0.3$, $z=-0.1$
7. $p-pq$; use $p=\frac{2}{3}$ and $q=-9$
8. $\dfrac{6+y-z}{3}$; use $y=3$, $z=9$
9. $c-(a-1)$; use $a=\dfrac{1}{2}$ and $c=\dfrac{3}{4}$
10. $\dfrac{3}{2}(j-6)-\dfrac{4}{5}(6+k)$; use $j=4$ and $k=-1$
11. $\dfrac{4-(m-p)}{2}+q$; use $m=2$, $p=6$, $q=-3$
12. $10g+4h$; use $g=-0.3$, $h=0.2$

Distribute the following.

13. $8(x-4)$
14. $-3(1+6x-4y)$
15. $0.3(8v+9)$
16. $-(-5+9a)$
17. $-(3m+n-1)$
18. $\dfrac{1}{4}(8n-2)$
19. $4\left(\dfrac{3}{8}-\dfrac{5}{6}p+\dfrac{2}{7}r\right)$
20. $-\dfrac{2}{5}(15a-20b+35)$

Combine like terms.

21. $-7x-x$
22. $(b+10)(-9)$
23. $-\dfrac{3}{5}x+\dfrac{2}{5}-\dfrac{4}{5}$
24. $x-10-6x+1$
25. $-\dfrac{5}{8}b+b$
26. $(5p-6)+(1+p)$

27. $-10\left(\dfrac{4}{5}(3-p)\right)$

28. $1-10n-10$

29. $-8x+(9+3x)$

30. $-6(5-m)+3m$

31. $1.5m-3.2-2.8m+5.6$

32. $3-10(x-2)$

33. $2(b+3)-(2b+1)$

34. $-\dfrac{2}{3}(6-9a)$

35. $4(x+7)+(x+4)$

36. $-10-4(n-5)$

37. $6(3+v)+9(4v-2)$

38. $\dfrac{5}{9}n-\dfrac{3}{8}+n+\dfrac{5}{4}$

39. $-5(2-6k)+10(1-3k)$

40. $2-7(4x+3)$

41. $-15\left(\dfrac{4x-2}{5}\right)+6\left(\dfrac{x+2}{2}\right)$

42. $14\left(\dfrac{3y-5}{7}\right)-8\left(\dfrac{2-y}{8}\right)$

1.3 Property of Equality

Objective: To use the property of equality to solve simple linear equations.

Now that we have learned about expressions, our next step is to work with equations. An **equation** is a mathematical statement containing the "equal" sign. It consists of two expressions, one on each side of the "equal" sign. Examples of equations are: $3x = 7$, $x + y = 2$, $6p + 5 = 2p - 1$.

A **linear equation in one variable** is an equation that can be written in the form $ax + b = c$, where $a, b,$ and c are real numbers and $a \neq 0$. It may also be referred to as a **first-degree equation**, because the exponent of the variable is understood to be 1. An example of linear equation in one variable is $7x + 2 = 10$.

A **solution** to a linear equation is a value that makes the equality true. For example, -5 is a solution to $4x + 16 = -4$ because if we replace x with -5 in the equation, we will get a true statement. Let us take a look at how to check a solution.

Check if $x = -5$ is a solution to $4x + 16 = -4$.

$$4(-5) + 16 \overset{?}{=} -4$$
$$-20 + 16 \overset{?}{=} -4$$
$$-4 = -4 \checkmark$$

Now the equation comes out to a true statement! Notice also that if another number was substituted, we would not get a true statement.

Check if $x = 3$ is a solution to $4x + 16 = -4$.

$$4(3) + 16 \overset{?}{=} -4$$
$$12 + 16 \overset{?}{=} -4$$
$$28 \neq -4$$

This demonstrates that $x = 3$ is not a solution to the equation.

World View Note The study of algebra was originally called the "Cossic Art" from the Latin, the study of "things" (which we now call variables). The father of algebra, Persian mathematician, Muhammad ibn Musa Khwarizmi, introduced the fundamental idea of balancing by subtracting the same quantity from both sides of the equation. He called the process al-jabr, which later became the word algebra.

Being able to solve linear equations and verify that we have the correct solution are important and fundamental skills in algebra. We will now learn a systematic approach to solving linear equations.

Consider a seesaw. If the weight on either end is equal, then the board will be level or balanced.

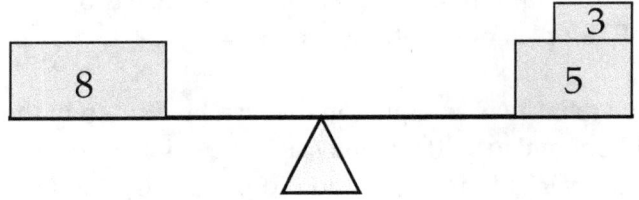

If we add (or remove) the same amount from both ends, the seesaw will still be balanced.

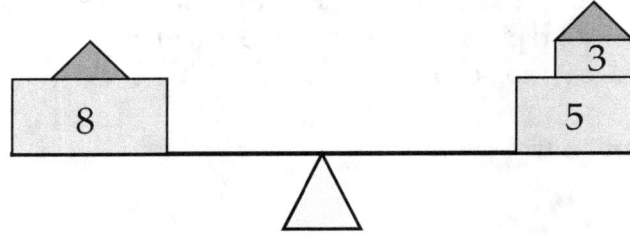

We will work with equations in the same way. Equations contain an equal symbol, =, indicating that the two sides of the equation must be equal or balanced. As long as we perform the same operations to each side of the equal sign, the equation will remain in balance.

Addition and Subtraction Property of Equality

Addition and Subtraction Property of Equality
If $a = b$, then $a + c = b + c$.
If $a = b$, then $a - c = b - c$.

This means that if two expressions are equal to each other, we can add or subtract the same number to each side of the equal sign and the two sides will remain equal. When we are solving a linear equation, our goal is to isolate the variable. That will give us the solution to the equation. Be sure to always check that you found the correct solution.

Example 1 Solve $x + 7 = -5$ for x. Be sure to check your answer.

Solution.

$$x + 7 = -5 \quad \text{Isolate the variable, x}$$
$$x + 7 - 7 = -5 - 7 \quad \text{Subtract 7 from each side of the equation}$$
$$x = -12 \quad \text{Our Solution}$$

1.3 Property of Equality

Check if $x = -12$ is the correct solution.

$(-12) + 7 \stackrel{?}{=} -5$ Substitute $x = -12$ into equation

$-5 = -5$ ✓   This verifies that $x = -12$ is the solution.

Exercise 1 Class Example
Solve the following equations. Be sure to check your answers.

a) $c + 7 = -3$ b) $5 = 8 + d$

Exercise 2 You Try
Solve the following equations. Be sure to check your answers.

a) $g + 4 = -8$ b) $9 + k = 7$

The same process is used in the following examples. Notice that this time we are isolating the variable by adding.

Example 2 Solve $x - 5 = 6$ for x. Be sure to check your answer.

Solution.

$$x - 5 = 6$$ Isolate the variable, x
$$x - 5 + 5 = 6 + 5$$ Add 5 to each side of the equation
$$x = 11$$ Our Solution

Check if $x = 11$ is the correct solution.

$$(11) - 5 \stackrel{?}{=} 6$$ Substitute $x = 11$ into the equation
$$6 = 6 \checkmark$$ This verifies that $x = 11$ is the solution.

Exercise 3 Class Example
Solve the following equations. Be sure to check your answers.

a) $w - 7 = -10$

b) $9 = -3 + y$

Exercise 4 You Try
Solve the following equations. Be sure to check your answers.

a) $m - 9 = 3$

c) $p - \frac{1}{2} = \frac{2}{3}$

b) $-4 + x = -7$

d) $1.5 = n - 2.3$

1.3 Property of Equality

Multiplication and Division Property of Equality

> **Multiplication Property of Equality**
> If $a = b$, then $ac = bc$.

This means that if we multiply each side of the equal sign by the same number, the two sides remain equal.

> **Division Property of Equality**
> If $a = b$, then $\dfrac{a}{c} = \dfrac{b}{c}$, for any $c \neq 0$.

This means that if we divide either side of the equal sign by the same number other than 0, the two sides remain equal.

With a multiplication problem, we can isolate the variable by dividing each side of the equation by the variable's coefficient. It is very important that care is taken with signs. If the variable's coefficient is a negative number, be sure to divide each side by the negative number.

Example 3 Solve $4m = -20$ for m. Be sure to check your answer.

Solution.

$\quad 4m = -20 \qquad$ Isolate the variable, m

$\quad \dfrac{4m}{4} = \dfrac{-20}{4} \qquad$ Divide each side by 4, the coefficient of m

$\quad m = -5 \qquad$ Our Solution

Check if $m = -5$ is the correct solution.

$\quad 4(-5) \stackrel{?}{=} -20 \qquad$ Substitute $m = -5$ into the equation

$\quad -20 = -20 \ \checkmark \qquad$ This verifies that $m = -5$ is the solution.

Example 4 Solve $-5x = 30$ for x. Be sure to check your answer.

Solution.

$\quad -5x = 30 \qquad$ Isolate the variable, x

$\quad \dfrac{-5x}{-5} = \dfrac{30}{-5} \qquad$ Divide each side by -5, the coefficient of x

$\quad x = -6 \qquad$ Our Solution

Check if $x = -6$ is the correct solution.

$$-5(-6) \stackrel{?}{=} 30 \qquad \text{Substitute } x = -6 \text{ into the equation}$$
$$30 = 30 \checkmark \qquad \text{This verifies that } x = -6 \text{ is the solution.}$$

Example 5 Solve $-x = 8$ for x. Be sure to check your answer.

Solution.
Note that we are solving for x. There is a negative sign in front of the variable, x, which needs to be cleared so that the variable is isolated.

Method 1:
$$-x = 8 \qquad \text{Isolate the variable x}$$
$$(-1)(-x) = (-1)(8) \qquad \text{Multiply each side by } -1$$
$$x = -8 \qquad \text{Our Solution}$$

Method 2:
$$-x = 8 \qquad \text{Isolate the variable x}$$
$$\frac{-x}{-1} = \frac{8}{-1} \qquad \text{Divide each side by } -1$$
$$x = -8 \qquad \text{Our Solution}$$

Both methods yield the same answer $x = -8$. Check if $x = -8$ is the correct solution.

$$-(-8) \stackrel{?}{=} 8 \qquad \text{Substitute } x = -8 \text{ into the equation}$$
$$8 = 8 \checkmark \qquad \text{This verifies that } x = -8 \text{ is the solution.}$$

1.3 Property of Equality

Exercise 5 Class Example
Solve the following equations. Be sure to check your answers.

a) $3m = -21$

b) $-x = -24$

c) $-6p = 20$

d) $4.2 = 7w$

Exercise 6 You Try
Solve the following equations. Be sure to check your answers.

a) $-2n = -20$

b) $-y = 5$

c) $4w = 1$

d) $-4.8 = 6v$

When the coefficient of the variable is a fraction, we can multiply by the reciprocal of the coefficient to isolate the variable.

Example 6 Solve $\frac{x}{5} = -3$ for x. Be sure to check your answer.

Solution.

$\frac{x}{5} = -3$ Coefficient of x is $\frac{1}{5}$

$(5)\frac{x}{5} = -3(5)$ Multiply each side by 5, the reciprocal of $\frac{1}{5}$

$x = -15$ Our Solution

Check if $x = -15$ is the correct solution.

$\frac{-15}{5} \stackrel{?}{=} -3$ Substitute $x = -15$ into the equation

$-3 = -3$ ✓   This verifies that $x = -15$ is the solution.

Exercise 7 Class Example
Solve the following equations. Be sure to check your answers.

a) $\frac{m}{7} = -2$ b) $-\frac{3}{5}p = 9$

Exercise 8 You Try
Solve the following equations. Be sure to check your answers.

a) $\frac{y}{2} = 8$ b) $\frac{2}{3}p = -24$

1.3 Property of Equality

The process described in this section is fundamental to solving equations. Once this process is mastered, we will see problems that involve several more steps. These problems may seem more complex, but the processes and patterns used will remain the same.

> **World View Note** French mathematician, Rene Descartes, wrote a book which included an appendix on geometry. It was in this book that he suggested using letters from the end of the alphabet for unknown values. This is why, in solving equations, we are often solving for the variables $x, y,$ and z.

1.3: Exercises

Solve each equation.

1. $v + 9 = 16$
2. $14 = -b$
3. $g - 1.1 = -1.6$
4. $-\dfrac{3}{5} = x - \dfrac{5}{4}$
5. $30.5 = a + 20.8$
6. $-1\dfrac{3}{7} + k = 5\dfrac{1}{2}$
7. $h - 7 = -26$
8. $-13 + p = -19$
9. $13 = n + 0.5$
10. $-22 = 16 + m$
11. $4r = -28$
12. $25 = -7 + x$
13. $-9 = \dfrac{n}{12}$
14. $\dfrac{5}{9} = \dfrac{y}{3}$
15. $0.2v = -16$
16. $-\dfrac{1}{3}x = -\dfrac{1}{9}$
17. $\dfrac{5}{8} = \dfrac{3}{4}n$
18. $\dfrac{1}{2} = 8a$
19. $\dfrac{5}{2}x = 25$
20. $\dfrac{k}{13} = -6$
21. $-16 + n = -13$
22. $-x = 21$
23. $15 = 12y$
24. $\dfrac{p}{20} = -2$
25. $\dfrac{r}{8} = \dfrac{5}{14}$
26. $p - \dfrac{3}{8} = -\dfrac{2}{5}$
27. $-10n = 35$
28. $\dfrac{5}{6} = 2c$

Rescue Roody!

29. Roody was told to solve $-9x = 5$. Roody got $x = 15$, but was marked incorrect. This is his work. Help Roody solve the equation.

$$\begin{aligned} -9x &= 6 \\ +9 & +9 \\ \hline x &= 15 \end{aligned}$$

1.4 Linear Equations - General Case

Objective: To solve linear equations

Now that we have learned how to solve one-step linear equations, we are ready to solve *any* kind of linear equation in one variable. As in the previous section, our goal is to isolate the variable of our equation. The basic idea to achieve this will be to use the properties covered in the previous section, one at a time.

Variable on One Side of the Equation

Example 1 Solve $3x + 4 = 13$ for x. Be sure to check your answer.

Solution.

$$3x + 4 = 13 \quad \text{Isolate the variable } x$$
$$3x + 4 - 4 = 13 - 4 \quad \text{Subtract 4 from each side}$$
$$3x = 9 \quad \text{Combine like terms}$$
$$\frac{3x}{3} = \frac{9}{3} \quad \text{Divide each side by 3}$$
$$x = 3 \quad \text{Our Solution}$$

Check if $x = 3$ is the correct solution.

$$3(3) + 4 \stackrel{?}{=} 13 \quad \text{Substitute } x = 3 \text{ into the equation}$$
$$9 + 4 = 13 \checkmark \quad \text{This verifies that } x = 3 \text{ is the solution.}$$

Exercise 1 Class Example
Solve $4x - 20 = 8$ for x. Be sure to check your answer.

Exercise 2 **You Try** Solve the following equations. Be sure to check your answers.

a) $5x - 3 = 12$
b) $4y + 2 = -10$

Variable on Each Side of the Equal Sign

If the variable appears on each side of the equal sign, our first goal is to take the equation to a form where the variable appears only on one side of the equation. To achieve this, we will use the properties covered in the previous section.

Example 2 Solve $3x - 4 = x + 3$ for x. Be sure to check your answer.

Solution.

$3x - 4 = x + 3$	Isolate the variable x
$3x + 4 - x = x + 3 - x$	Subtract x from each side
$2x + 4 = 3$	Combine like terms
$2x + 4 - 4 = 3 - 4$	Subtract 4 from each side
$2x = 7$	Combine like terms
$\dfrac{2x}{2} = \dfrac{7}{2}$	Divide each side by 2
$x = \dfrac{7}{2}$	Our Solution

1.4 Linear Equations - General Case

Check if $x = \dfrac{7}{2}$ is the correct solution.

$3\left(\dfrac{7}{2}\right) - 4 \stackrel{?}{=} \left(\dfrac{7}{2}\right) + 3$ Substitute $x = \dfrac{7}{2}$ into the equation

$\dfrac{21}{2} - 4 \stackrel{?}{=} \dfrac{7}{2} + 3$ Find LCD

$\dfrac{21}{2} - \dfrac{8}{2} \stackrel{?}{=} \dfrac{7}{2} + \dfrac{6}{2}$ Perform indicated operation

$\dfrac{13}{2} = \dfrac{13}{2}$ ✓ This verifies that $x = \dfrac{7}{2}$ is the solution.

Exercise 3 Class Example

Solve $4y - 3 = 2y - 1$ for y. Be sure to check your answers.

Exercise 4 You Try

Solve the following equations. Be sure to check your answers.

a) $5x + 1 = 12 - x$

b) $4y + 2 = 10y + 1$

Example 3 Solve $3(x-2) = 4(x+1)$ for x. Be sure to check your answer.

Solution.

$3(x-2) = 4(x+1)$	Distribute
$3x - 6 = 4x + 4$	Isolate the variable x
$3x - 6 - 4x = 4x + 4 - 4x$	Subtract 4x from each side
$-x - 6 = 4$	Combine like terms
$-x - 6 + 6 = 4 + 6$	Add 6 to each side
$-x = 10$	Note: x is not yet completely isolated
$(-1)(-x) = (-1)(10)$	Multiply each side by -1
$x = -10$	Our Solution

Check if $x = -10$ is the correct solution.

$3((-10) - 2) \stackrel{?}{=} 4((-10) + 1)$	Substitute $x = -10$ into the equation
$3(-12) \stackrel{?}{=} 4(-9)$	Perform indicated operation inside parenthesis
$-36 = -36$ ✓	This verifies that $x = -10$ is the solution.

Exercise 5 Class Example
Solve the following equations. Be sure to check your answers.

a) $3(2p + 8) = p + 24$
b) $5(3 - n) + 2 = 3(n + 2)$

1.4 Linear Equations - General Case

Exercise 6 You Try
Solve the following equations. Be sure to check your answers.

a) $5(x+2) = 3(x-3)$

b) $4(y+3) - 2(y-1) = 0$

Example 4 Solve $3(x-2) - x = 2x + 1$ for x.

Solution.

$3(x-2) - x = 2x+1$	Distribute the 3
$3x - 6 - x = 2x + 1$	Combine like terms
$2x - 6 = 2x + 1$	Move variable x to one side of equation
$2x - 6 - 2x = 2x + 1 - 2x$	Subtract 2x from each side
$-6 = 1$	This is clearly false!!

If we reach a false statement, such as $-6 = 1$, this says that there is no value of x that will make the equality true. This means that our equation has **No Solution**.

Example 5 Solve $4(x+3) = 2x + 2(x+6)$ for x.

Solution.

$4(x+3) = 2x + 2(x+6)$	Distribute 4 and 2
$4x + 12 = 2x + 2x + 12$	Combine like terms
$4x + 12 = 4x + 12$	we have the same expression on each side!!

Something interesting happened in this example. We see that our last line is an equality that will be true for **any value of** x. This means that any real number is a solution to this equation. We will express this by saying that the solution consists of **all real numbers, ℝ**.
Notice that if we continue working to try to isolate the variable in our last example, we obtain the

following.

$$4x+12 = 4x+12 \qquad \text{Isolate the variable, x}$$
$$4x+12-4x = 4x+12-4x \qquad \text{Subtract 4x from each side}$$
$$12 = 12 \qquad \text{A true statement!!}$$

This last statement, $12 = 12$, is true, regardless of the value of x. This tells us that the solution to our original equation is the set of all real numbers, $\mathbb{R}$.

When solving a linear equation in one variable, we may encounter any one of the following cases.

- We are able to isolate the variable and get a numerical solution. The solution is verified to be the correct solution when substituted back into the equation.

- We obtain a false statement such as $-6 = 11$. In this case, the equation is said to have **No Solution**.

- We obtain a true statement, regardless of the value of our variable, such as $12 = 12$. The solution to the equation consists of **All Real Numbers, $\mathbb{R}$**.

Exercise 7 **Class Example**
Solve the following equations. Be sure to check your answers, if applicable.

a) $7x + 4(2-x) = 3(x+4)$

b) $5(y-2) - (y+1) = 4y - 11$

1.4 Linear Equations - General Case

Exercise 8 You Try

Solve the following equations. Be sure to check your answers, if applicable.

a) $5(m+2)+1=2(m-3)+3m$
b) $4(1-p)+4(p-1)=0$

Equations with Fractions

The next examples show how to solve equations involving fractions. Two different methods are shown.

Example 6 Solve $\frac{1}{2}x+3=\frac{3}{4}$ for x. Be sure to check your answer.

Solution.

Method 1: Work directly with the fractions.

$$\frac{1}{2}x+3=\frac{3}{4} \qquad \text{Isolate the variable, x}$$

$$\frac{1}{2}x+3-3==\frac{3}{4}-3 \qquad \text{Subtract 3 from each side}$$

$$\frac{1}{2}x=\frac{3}{4}-\frac{12}{4} \qquad \text{Find LCD}$$

$$\frac{1}{2}x=-\frac{9}{4} \qquad \text{Subtract}$$

$$2\left(\frac{1}{2}x\right)=2\left(-\frac{9}{4}\right) \qquad \text{Multiply each side by 2, the reciprocal of } \frac{1}{2}$$

$$x=-\frac{9}{2} \qquad \text{Our Solution}$$

Method 2: Multiply each side by the LCD of all fractions appearing in the equation.

$$\frac{1}{2}x + 3 = \frac{3}{4}$$ Isolate the variable, x

$$4\left(\frac{1}{2}x + 3\right) = 4\left(\frac{3}{4}\right)$$ Multiply each side by the LCD, 4 and Distribute

$$2x + 12 = 3$$
$$2x + 12 - 12 = 3 - 12$$ Subtract 12 from each side
$$2x = -9$$ Combine like terms
$$\frac{2x}{2} = -\frac{9}{2}$$ Divide each side by 2
$$x = -\frac{9}{2}$$ Our Solution

Check if $x = -\frac{9}{2}$ is the correct solution.

$$\frac{1}{2}\left(-\frac{9}{2}\right) + 3 \stackrel{?}{=} \frac{3}{4}$$ Substitute $x = -\frac{9}{2}$ into equation

$$-\frac{9}{4} + 3 \stackrel{?}{=} \frac{3}{4}$$ Multiply $\frac{1}{2}\left(-\frac{9}{2}\right)$

$$-\frac{9}{4} + \frac{12}{4} \stackrel{?}{=} \frac{3}{4}$$ Find LCD and add

$$\frac{3}{4} = \frac{3}{4} \checkmark$$ This verifies that $x = -\frac{9}{2}$ is the solution.

1.4 Linear Equations - General Case

Example 7 Solve $\frac{3}{4}(x-2) = x + \frac{5}{2}$ for x. Be sure to check your answer.

Solution.

Method 1: Work directly with the fractions.

$$\frac{3}{4}(x-2) = x + \frac{5}{2}$$

$\frac{3x}{4} - \frac{3}{2} = x + \frac{5}{2}$ Distribute $\frac{3}{4}$

$\frac{3x}{4} - \frac{3}{2} + \frac{3}{2} = x + \frac{5}{2} + \frac{3}{2}$ Add $\frac{3}{2}$ to each side

$\frac{3x}{4} = x + \frac{8}{2}$ Combine like terms

$\frac{3x}{4} = x + 4$ Simplify $\frac{8}{2}$

$\frac{3x}{4} - x = x + 4 - x$ Subtract x from each side

$\frac{3x}{4} - x = 4$ Combine like terms

$\frac{3x}{4} - \frac{4x}{4} = 4$ Subtract by finding LCD

$\frac{-x}{4} = 4$ Combine like terms

$(-4)\left(\frac{-x}{4}\right) = (-4)(4)$ Multiply each side by -4, the reciprocal of $\frac{-1}{4}$

$x = -16$ Our Solution

Method 2: Multiply each side by the LCD of all fractions appearing in the equation.

$$\frac{3}{4}(x-2) = x + \frac{5}{2}$$

$$4\left(\frac{3}{4}(x-2)\right) = 4\left(x + \frac{5}{2}\right) \quad \text{Multiply each side by the LCD, 4}$$

$$3(x-2) = 4\left(x + \frac{5}{2}\right) \quad \text{Multiply } 4\left(\frac{3}{4}\right) \text{ using associative property}$$

$$3x - 6 = 4x + 10 \quad \text{Distribute 3 and 4}$$

$$3x - 6 - 3x = 4x + 10 - 3x \quad \text{Subtract } 3x \text{ from each side}$$

$$-6 = x + 10 \quad \text{Combine like terms}$$

$$-6 - 10 = x + 10 - 10 \quad \text{Subtract 10 from each side}$$

$$-16 = x \quad \text{Our Solution}$$

Check if $x = -16$ is the correct solution.

$$\frac{3}{4}(-16 - 2) \stackrel{?}{=} -16 + \frac{5}{2} \quad \text{Substitute } x = -16 \text{ into equation}$$

$$\frac{3}{4}(-18) \stackrel{?}{=} \frac{-32}{2} + \frac{5}{2} \quad \text{Left: subtract; Right: find LCD}$$

$$\frac{-27}{2} = \frac{-27}{2} \checkmark \quad \text{This verifies that } x = -16 \text{ is the solution.}$$

Exercise 9 Class Example
Solve the following equations. Be sure to check your answers.

a) $\frac{5}{6}y - 2 = \frac{2}{3}$

b) $\frac{3}{5}(x - 3) = 4x + \frac{1}{4}$

1.4 Linear Equations - General Case

Exercise 10 You Try

Solve the following equations. Be sure to check your answers.

a) $5 + \dfrac{2}{3}x = \dfrac{1}{2}x$

b) $\dfrac{3}{2}(x-2) = \dfrac{x-2}{3}$

1.4: Exercises

Solve each equation.

1. $2a - 3 = 9$
2. $5x - 12 = 8$
3. $3y + 12 = 0$
4. $24 = 2n - 8$
5. $2 = 2a - 5$
6. $14k + 22 = 1$
7. $4y + 7 = 7y - 8$
8. $5 - 2p = p - 4$
9. $5x + 9 = 21 - 5x$
10. $a + 5 = 5 - 5a$
11. $5n - 14 = 9n + 4$
12. $5 - 2(p + 3) = 4p + 7$
13. $5y + 34 = -2(1 - 7y)$
14. $12 - 4x = 8(3x + 1)$
15. $4(6a + 1) = 43 + 3(4a + 3)$
16. $36 - 6(7 + y) = 2(1 + 7y)$
17. $2(4 - 8n) = 8(1 - n)$
18. $7(k - 4) + 9 = 12 + 3(2k - 9)$
19. $3(x - 2) + 6 = 4(2x - 1) - 15$
20. $7(m - 2) = 6(m - 1) + 4$
21. $4(1 + a) + 8(5 + 3a) = 2a$
22. $5(y + 7) - 4(8y - 2) = 7$
23. $5 + \frac{1}{4}n = 4$
24. $\frac{3}{5}(1 + p) = \frac{21}{20}$
25. $\frac{5}{6} + \frac{1}{2}(m + 4) = \frac{2}{3}$
26. $2x + \frac{9}{5} = -\frac{11}{5}$
27. $\frac{1}{3}(n - 12) = \frac{1}{4}(n + 8)$
28. $\frac{1}{4}(4a - 1) + \frac{25}{6} = -\frac{1}{3}$

1.4 Linear Equations - General Case

Mid-Chapter 1 Check-Up

Simplify the following expressions.

1. $-2(6-9)^2 - 5$

3. $3(4k-1) - 2(6k+4)$

2. $\dfrac{-12 - 6^2}{-15 + 5}$

4. $5 - \dfrac{3}{4}(12n - 8)$

Evaluate each expression using the values given.

5. $a^2 - b$; use $a = -3$, $b = -6$

6. $\dfrac{8 + 2m}{n^2 + 1}$; use $m = -4$, $n = 2$

Solve the following equations.

7. $m + 7 = -2$

10. $2x - 8 = 6x + 12$

8. $\dfrac{m}{3} = 9$

11. $2(9 - y) = y - 3(y - 6)$

9. $\dfrac{2}{5}m + 10 = 0$

12. $\dfrac{1}{2}(6 - m) = \dfrac{1}{3}(2m + 9)$

1.5 Linear Inequalities

Objective: To solve, graph, and give interval notation for the solution to linear inequalities

When we have an equation such as $x = 4$, there is a specific value assigned to the variable, x. With inequalities, a range of values is assigned to the variable. The following symbols are used in inequalities:

>   Greater than
⩾   Greater than or equal to
<   Less than
⩽   Less than or equal to

An inequality such as $x < 4$ means that our variable can be any number smaller than 4 such as $-2, 0, 3, -\frac{3}{5}, 3.9$, or even 3.99999999 while an inequality such as $x \geqslant -2$ means that our variable can be any number greater than or equal to -2 such as $5, 0, -\frac{7}{4}, -1.999999$, including -2.

Because we do not have one set value for our variable, it is often useful to draw a picture of the solutions on a number line. Once the graph is drawn, we will also write our solution in interval notation. Interval notation gives two numbers, the first is the solution's smallest value and the second is the solution's largest value. If the solution has no largest value, we will use ∞ (infinity). If the solution has no smallest value, we will use $-\infty$ (negative infinity).

To graph on the number line, inequalities that do not have an equal sign will use open circles for graphs and parenthesis for interval notation. Inequalities that involve the equal sign will use closed circles for graphs and brackets for interval notation. The number line will be shaded in the appropriate region to denote all possible solutions of the variable.

Note. Infinity is not a number. It is an idea of something without end. When writing your solution in interval notation, **infinity will always be enclosed by a parenthesis, not a bracket**.

World View Note The English mathematician and astronomer, Thomas Harriot, first introduced the inequality symbols. It is said that while Harriot was surveying North America, he saw a native American with the symbol, ⋈ on his arm. It is likely that Harriot developed the inequality symbol from this.

Example 1 Graph $x < 2$ on the number line and write your solution in interval notation.

Solution.
We want all solutions of x that are smaller than 2, but not including 2. On the number line, we

will use an open circle on 2, to show that 2 is not included and shade everything to the left of 2 to show all solutions less than 2.

Interval notation is $(-\infty, 2)$ since there is no smallest value and the largest value is 2, excluding 2.

Example 2 Graph $y \geq -1$ on the number line and write your solution in interval notation.

Solution.
We want all solutions of y that are greater than -1, including -1. On the number line, we will use a closed circle around -1 to show that -1 is included and shade everything to the right of -1 to show all solutions greater than -1.

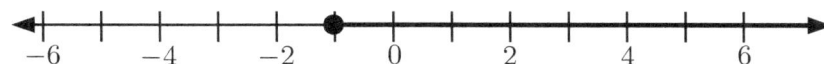

Interval notation is $[-1, \infty)$ since the smallest value is -1 and there is no largest value.

Exercise 1 **Class Example**

a) Graph $p \leq 5$ on the number line and write your solution in interval notation.

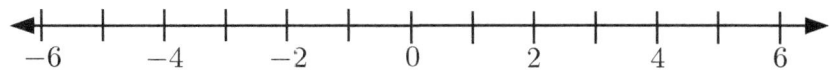

b) Graph $-3 > z$ on the number line and write your solution in interval notation

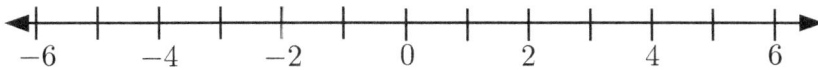

It is important to be careful when the inequality is written backwards as in the previous example. Students often draw their graphs the wrong way when this is the case. One way to avoid graphing the wrong way is to rewrite the inequality. For example, rewrite $-3 > z$ as $z < -3$.

Exercise 2 **You Try**

a) Graph $n \leq 0$ on the number line and write your solution in interval notation.

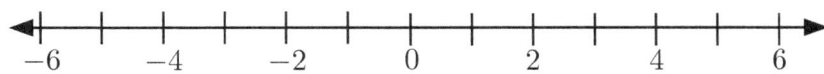

b) Graph $-4 < c$ on the number line and write your solution in interval notation

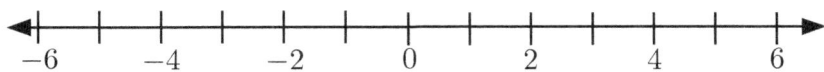

Example 3 Find the inequality that represents the graph shown and write the solution in interval notation.

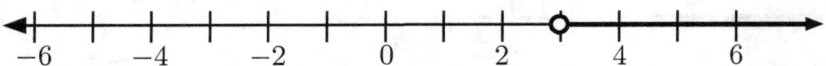

Solution.
There is an open circle on 3 and the shaded portion is to the right of 3. The solution must be everything greater than 3 but not including 3. The solution in inequality notation is $x > 3$ and in interval notation is $(3, \infty)$.

Example 4 Find the inequality that represents the graph shown and write the solution in interval notation.

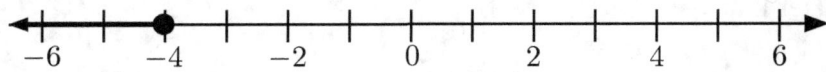

Solution.
There is a closed circle on -4 and the shaded portion is to the left of -4. The solution must be everything less than -4, including -4. The solution in inequality notation is $x \leqslant -4$ and in interval notation is $(-\infty, -4]$.

Exercise 3 You Try
Find the inequality that represents each graph below and write the solution in interval notation.

a)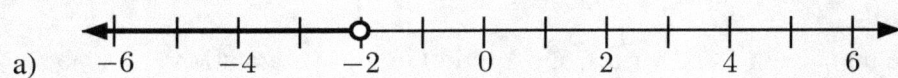

b)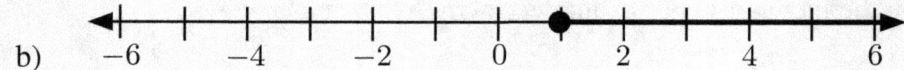

1.5 Linear Inequalities

Solving Linear Inequalities

Solving inequalities is very similar to solving equations. Consider the following inequalities. Notice what happens to the inequality sign as we add, subtract, multiply and divide by both positive and negative numbers to keep the statement a true statement.

Start with	*Operation Performed*	*Result*	*Inequality Sign*
$6 > 2$	Add 2 to each side	$8 > 4$	Correct
$6 > 2$	Subtract 5 from each side	$1 > -3$	Correct
$6 > 2$	Add -4 to each side	$2 > -2$	Correct
$6 > 2$	Subtract -8 from each side	$14 > 10$	Correct
$6 > 2$	Multiply each side by 7	$42 > 14$	Correct
$6 > 2$	Divide each side by 6	$1 > \frac{1}{3}$	Correct
$6 > 2$	Multiply each side by -3	$-18 > -6$	***Incorrect***
$6 > 2$	Divide each side by -1	$-6 > -2$	***Incorrect***

As illustrated above, the inequality is preserved when we add or subtract to each side of the inequality. The same is true when we multiply or divide by a positive number. But if we multiply or divide by a negative number, the inequality reverses directions.

Example 5 Solve the inequality $y - 3 < -5$ for y. Graph on a number line and write the solution in interval notation.

Solution.

$$y - 3 < -5 \qquad \text{Isolate the variable, } y$$
$$y - 3 + 3 < -5 + 3 \qquad \text{Add 3 to each side of the inequality}$$
$$y < -2 \qquad \text{Our Solution in inequality notation}$$
$$(-\infty, -2) \qquad \text{Our Solution in interval notation}$$

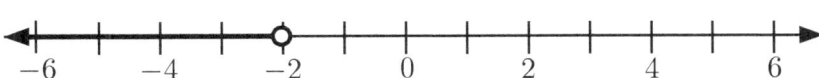

Example 6 Solve the inequality $-5m \geq 20$ for m. Graph on a number line and write the solution in interval notation.

Solution.

$$-5m \geq 20 \qquad \text{Isolate the variable, y}$$
$$\frac{-5m}{-5} \leq \frac{20}{-5} \qquad \text{Divide each side by } -5; \text{ Reverse inequality symbol}$$
$$m \leq -4 \qquad \text{Our Solution in inequality notation}$$
$$(-\infty, -4] \qquad \text{Our Solution in interval notation}$$

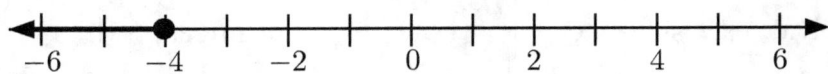

Exercise 4 Class Example
Solve the following inequality. Graph on a number line and write your solution in interval notation.

a) $9 + m \geq 6$

b) $4 > \dfrac{n}{6}$

1.5 Linear Inequalities

Exercise 5 You Try

Solve the following inequality. Graph on a number line and write your solution in interval notation.

a) $g + \dfrac{2}{3} \leq \dfrac{1}{5}$

b) $5 > -\dfrac{1}{2}k$

Example 7 Solve the inequality $5 - 2x \geq 11$ for x. Graph on a number line and write your solution in interval notation.

Solution.

$$5 - 2x \geq 11 \qquad \text{Isolate the variable, x}$$
$$5 - 2x - 5 \geq 11 - 5 \qquad \text{Subtract 5 from each side}$$
$$-2x \geq 6 \qquad \text{Divide each side by } -2$$
$$\dfrac{-2x}{-2} \leq \dfrac{6}{-2} \qquad \text{Reverse inequality}$$
$$x \leq -3 \qquad \text{Our Solution in inequality notation}$$
$$(-\infty, -3] \qquad \text{Our Solution in interval notation}$$

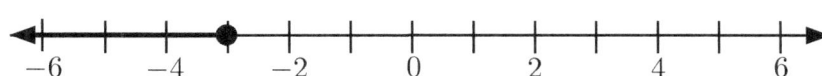

Example 8 Solve the inequality $11x - 12 < 4(3x - 6) + 8$ for x. Graph on a number line and write your solution in interval notation.

Solution.

$$11x - 12 < 4(3x - 6) + 8 \quad \text{Distribute 4}$$
$$11x - 12 < 12x - 24 + 8 \quad \text{Combine like terms}$$
$$11x - 12 < 12x - 16 \quad \text{Subtract 12x from each side}$$
$$11x - 12 - 12x < 12x - 16 - 12x \quad \text{Combine like terms}$$
$$-x - 12 < -16 \quad \text{Add 12 to each side}$$
$$-x - 12 + 12 < -16 + 12 \quad \text{Combine like terms}$$
$$-x < -4 \quad \text{Divide each side by } -1$$
$$\frac{-x}{-1} > \frac{-4}{-1} \quad \text{Reverse inequality}$$
$$x > 4 \quad \text{Our Solution in inequality notation}$$
$$(4, \infty) \quad \text{Our Solution in interval notation}$$

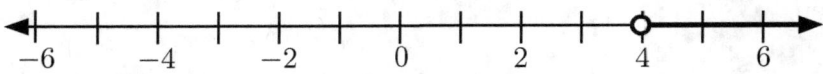

Exercise 6 Class Example
Solve the following inequality. Graph on a number line and write your solution in interval notation.

a) $4 - (g + 3) > 5g$

b) $6.8p + 4.2 \leqslant 3.5p - 12.3$

1.5 Linear Inequalities

Exercise 7 You Try
Solve the following inequality. Graph on a number line and write your solution in interval notation.

a) $3(y-2) \geq -(y+6)$

b) $\frac{4}{5}w - \frac{1}{2} > \frac{3}{10}w + 2$

1.5: Exercises

Complete the table below.

	Inequality Notation	Number Line Graph	Interval Notation
1.	$n > -5$		
2.			$[4, \infty)$
3.		open circle at -2, shaded left	
4.	$1 \geq k$		
5.			$(-\infty, -5)$
6.		closed circle at 1, shaded left	
7.	$-6 \leq p$		
8.		open circle at -1, shaded right	
9.			$(-\infty, 5]$
10.	$x < 4$		

Solve each inequality, graph on a number line, and write the solution in interval notation.

11. $2 + r < 3$

12. $-7n - 10 \geq 60$

13. $-8(n - 5) \geq 0$

14. $8 + \dfrac{n}{3} > 6$

15. $\dfrac{6 + x}{12} \leq -1$

16. $24 \geq -6(m - 6)$

17. $-2(3 + k) < -44$

18. $-r - 5(r - 6) < -18$

19. $11 \geq 8 + \dfrac{x}{2}$

20. $24 + 4b < 4(1 + 6b)$

21. $-8(2 - 2n) \geq -16 + n$

22. $4 + 2(a + 5) < -2(-a - 4)$

23. $-(k - 2) > -k - 20$

24. $-36 + 6x > -8(x + 2) + 4x$

25. $-5v - 5 \leq -5(4v + 1)$

Rescue Roody!

26. Roody was told to solve the inequality $6 - 2(m + 3) \geq 7 - 2m$, but he keeps getting stuck. Help

1.5 Linear Inequalities

Roody. Here is his work.

$$6 - 2(m+3) \geqslant 7 - 2m$$
$$6 - 2m - 6 \geqslant 7 - 2m$$
$$-2m \geqslant 7 - 2m$$
$$\underline{+2m \quad +2m}$$
$$0m \geqslant 7$$

1.6 Absolute Value Equations

Objective: To solve linear absolute value equations

Recall that the absolute value of a number is its distance from 0 on the number line. For example, $|3| = 3$ and $|-3| = 3$ since the numbers 3 and -3 are 3 units from 0 on the number line.

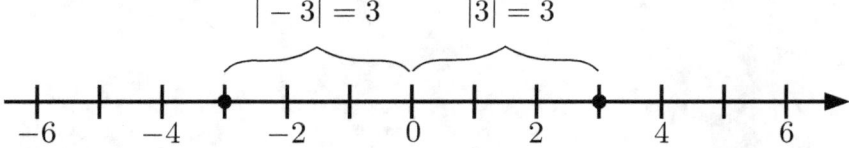

When dealing with equations involving absolute value, there can be more than one solution because the inside of the absolute value could be positive or negative. We must account for both possibilities.

Example 1 Solve $|x| = 3$ for x. Be sure to check your answers.

Solution.

$$|x| = 3 \qquad \text{x is 3 units from 0 on number line}$$
$$x = -3 \text{ or } x = 3 \qquad \text{Our Solutions}$$

Check if $x = -3$ is the correct solution.

$$|-3| \stackrel{?}{=} 3 \qquad \text{Substitute } x = -3 \text{ into the equation}$$
$$3 = 3 \checkmark \qquad \text{This verifies that } x = -3 \text{ is the solution.}$$

Check if $x = 3$ is the correct solution.

$$|3| \stackrel{?}{=} 3 \qquad \text{Substitute } x = 3 \text{ into the equation}$$
$$3 = 3 \checkmark \qquad \text{This verifies that } x = 3 \text{ is the solution.}$$

Notice that we have considered two possibilities as illustrated on the number line below.

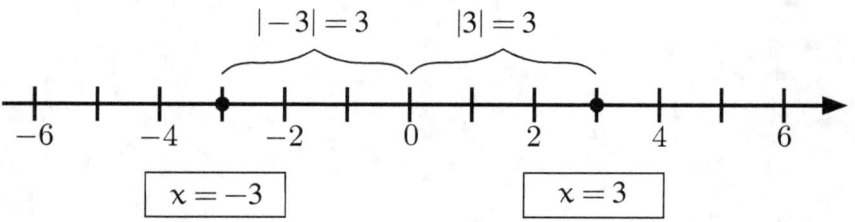

1.6 Absolute Value Equations

Exercise 1 Class Example
Solve each of the following equations. Be sure to check your answers.

a) $|x| = 7$

b) $|w| = 1.5$

Exercise 2 You Try
Solve each of the following equations. Be sure to check your answers.

a) $|x| = 9$

c) $|m| = 3.6$

b) $|y| = \dfrac{5}{2}$

d) $|n| = 0$

Often, we will have a more complex expression within the absolute value symbol. We still consider the distance the expression is from 0 on the number line. This is illustrated below on the number line given the problem: Solve $|\Delta| = a$, for $a > 0$.

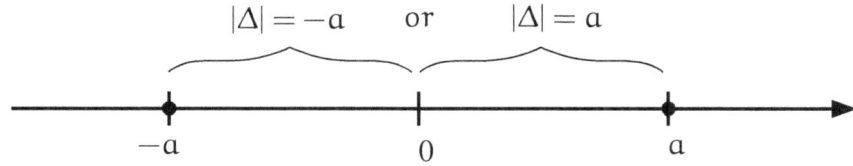

Example 2 Solve $|3x| = 12$ for x. Be sure to check your answers.

Solution.
The expression 3x is 12 units from 0 on the number line. This results in the following two linear equations.

$3x = -12$	or	$3x = 12$	Divide each equation by 3
$x = -4$	or	$x = 4$	Our Solutions

Check if $x = -4$ is a solution.

$$|3(-4)| \stackrel{?}{=} 12$$ Substitute $x = -4$ into equation
$$|-12| = 12 \checkmark$$ This verifies $x = -4$ is a solution.

Check if $x = 4$ is a solution.

$$|3(4)| \stackrel{?}{=} 12$$ Substitute $x = 4$ into equation
$$|12| = 12 \checkmark$$ This verifies $x = 4$ is a solution.

Example 3 Solve $|2x - 1| = 7$ for x. Be sure to check your answers.

Solution.
The expression $2x - 1$ is 7 units from 0 on the number line. This results in the following two linear equations.

$2x - 1 = -7$	or	$2x - 1 = 7$	Add 1 to each side of the equation
$2x = -6$	or	$2x = 8$	Divide each side by 2
$x = -3$	or	$x = 4$	Our Solution

Check if $x = -3$ is a solution.

$$|2(-3) - 1| \stackrel{?}{=} 7$$ Substitute $x = -3$ into equation
$$|-6 - 1| \stackrel{?}{=} 7$$ Follow order of operations
$$|-7| = 7 \checkmark$$ This verifies $x = -3$ is a solution.

Check if $x = 4$ is a solution.

$$|2(4) - 1| \stackrel{?}{=} 7$$ Substitute $x = 4$ into the equation
$$|8 - 1| \stackrel{?}{=} 7$$ Follow order of operations
$$|7| = 7 \checkmark$$ This verifies $x = 4$ is a solution.

1.6 Absolute Value Equations

Exercise 3 Class Example
Solve each of the following equations. Be sure to check your answers.

a) $|7x| = 21$

b) $|2y - 1| = 3$

Exercise 4 You Try
Solve each of the following equations. Be sure to check your answers.

a) $|v - 3| = 3$

c) $|4t - 3| = 11$

b) $|-x| = 9$

d) $|1 - p| = 0$

Remember the value of the absolute value must always be non-negative. Notice what happens in the next example.

Example 4 Solve $|x| = -2$ for x.

Solution.
Since the absolute value of any number cannot be negative, this equation does not have a solution! Our answer, therefore, is **no solution**.

Note. If you think the solution is $x = -2$ or $x = 2$, check your answer by substituting the value of x into the equation, $|x| = -2$. Neither answer will work.

$$\text{Substitute } x = -2 \qquad\qquad \text{Substitute } x = 2$$
$$|-2| \stackrel{?}{=} -2 \qquad\qquad\qquad |2| \stackrel{?}{=} -2$$
$$2 \neq -2 \qquad\qquad\qquad\quad 2 \neq -2$$

Exercise 5 Class Example
Solve the following equations.

a) $|w| = -7$

b) $|3x + 1| = -1$

Exercise 6 You Try
Solve the following equations.

a) $|x| = -9$

c) $|2p - 6| = -1$

b) $|1 - y| = -8$

d) $|x - 5| = 2$

1.6: Exercises

Solve each equation for the variable.

1. $|x| = 3$
2. $|y| = 0$
3. $|k| = -1$
4. $|5p| = 25$
5. $|9n + 8| = 46$
6. $|6 - 2c| = 24$
7. $|-3w| = 6$
8. $\left|\dfrac{4b + 10}{8}\right| = 3$
9. $|8(x + 7)| = 8$
10. $|y + 1| = 8$
11. $\left|\dfrac{3m + 1}{2}\right| = 5$
12. $\left|\dfrac{2}{3}x\right| = \dfrac{38}{9}$
13. $|4 + x| = -9$
14. $|3 - 4t| = 0$
15. $|-2g| = 10$
16. $|8 - k| = 8$
17. $|4c - 3| = 7$
18. $|m + 5| = -8$

1.7 Formulas

Objective: To solve for a variable in a given formula

A formula establishes a relationship between two or more variables. Sometimes a formula must be rewritten in a certain way that involves solving for a specific variable. Solving for a specific variable is much like solving general linear equations. The only difference is that we will have several variables in the problem and we will need to solve for one specific variable.

> **Example 1** The distance traveled, d, at rate, r, for a certain amount of time, t, is given by the formula, $d = rt$.
>
> a) Solve the formula for t.
>
> **Solution.**
>
> $$d = rt \qquad \text{Isolate the variable t}$$
> $$\frac{d}{r} = \frac{rt}{r} \qquad \text{Divide each side by r}$$
> $$\frac{d}{r} = t \qquad \text{Our Solution}$$
>
> b) Bert wants to see a concert 60 miles from his house. He expects to travel at a constant speed of 40 mph (miles per hour). How long will Bert be traveling?
>
> **Solution.**
>
> Since we already solved for t above, we can use that formula to find how long Bert will be traveling by substituting $d = 60$ miles and $r = 40$ mph.
>
> $$t = \frac{d}{r}$$
> $$t = \frac{60 \text{ miles}}{40 \text{ mph}} \qquad \text{Substitute d with 60 miles and r with 40 mph}$$
> $$t = 1.5 \text{ hours} \qquad \text{Our Solution}$$

The formula $m + n = p$ is solved for p. If we want to solve for n instead, we need to isolate the n on one side all by itself, with all the other terms on the other side of the equation.

1.7 Formulas

Example 2 Solve $m+n=p$ for n.

Solution.

$$m+n=p \qquad \text{Isolate the variable } n$$
$$m+n-m=p-m \qquad \text{Subtract } m \text{ from each side}$$
$$n=p-m \qquad \text{Our Solution}$$

As p and m are not like terms, they cannot be combined. For this reason, we leave the expression as $p-m$.

It is important to note that we have completed the problem when the variable we are solving for is isolated or alone on one side of the equation and **it does not appear anywhere on the other side of the equation**.

Exercise 1 Class Example
Solve the following formula for the indicated variable.

a) $A=lw$ for w

b) $x+y+z=180$ for z

Exercise 2 You Try
Solve the following equations.

a) $a-b=c$ for a

b) $C=2\pi r$ for r

Example 3 The perimeter, P, of a rectangle is $P = 2w + 2l$ where w is the width and l is the length of the rectangle.

a) Solve the formula for w.

Solution.

$$P = 2w + 2l \qquad \text{Isolate the variable } w$$
$$P - 2l = 2w + 2l - 2l \qquad \text{Subtract } 2l \text{ on each side}$$
$$\frac{P - 2l}{2} = \frac{2w}{2} \qquad \text{Divide each side by 2}$$
$$\frac{P - 2l}{2} = w \qquad \text{Our Solution}$$

Note. The solution can be rewritten by dividing **each** term in the numerator by the denominator as follows.

$$w = \frac{P - 2l}{2}$$
$$= \frac{P}{2} - \frac{2l}{2}$$
$$= \frac{P}{2} - l$$

It is incorrect to simplify only one term in the numerator (that is, divide $2l$ by 2 but not P to get $w = P - l$)

b) Suppose a rectangle is 8 inches long and has a perimeter 40 inches. Find the width of the rectangle.

Solution.

Since we already solved for w above, we can use that formula to find the rectangle's width by substituting $P = 40$ inches and $l = 8$ inches.

$$w = \frac{P - 2l}{2}$$
$$= \frac{40 - 2(8)}{2}$$
$$= \frac{40 - 16}{2}$$
$$= \frac{24}{2}$$
$$= 12 \qquad \text{So the rectangle is 12 inches wide.}$$

1.7 Formulas

Exercise 3 Class Example
Given the formula $6x - 3y = 24$.

a) Solve the formula for y.

b) Suppose $x = \dfrac{2}{3}$, find the value of y.

Exercise 4 You Try
Given the formula $3x + 2y = 12$.

a) Solve the formula for x.

b) Suppose $y = -\dfrac{3}{2}$, find the value of x.

Formulas with Parenthesis

We will show two methods to solve formulas with parenthesis. The answers may look different. With some algebraic manipulation, we can show that two different looking solutions are in fact the same.

Example 4 Solve $5(x - y) = b$ for x.

Solution.

Method 1: Keep the parenthesis as a group

$$5(x-y) = b \qquad \text{Keep } (x-y) \text{ as a group}$$

$$\frac{5(x-y)}{5} = \frac{b}{5} \qquad \text{Divide each side by 5}$$

$$x - y = \frac{b}{5}$$

$$x - y + y = \frac{b}{5} + y \qquad \text{Add } y \text{ to each side}$$

$$x = \frac{b}{5} + y \qquad \text{Our Solution}$$

Method 2: Use the Distributive Property

$$5(x-y) = b \qquad \text{Distribute 5}$$
$$5x - 5y = b \qquad \text{Isolate the variable, } x$$
$$5x - 5y + 5y = b + 5y \qquad \text{Add } 5y \text{ to each side}$$
$$5x = b + 5y$$

$$\frac{5x}{5} = \frac{b+5y}{5} \qquad \text{Divide each side by 5}$$

$$x = \frac{b+5y}{5} \qquad \text{Our Solution}$$

Be very careful as we isolate x so that we do not cancel the 5 in the numerator and the denominator of the fraction. This is not allowed as the 5 in the numerator is only in one of the terms. To do the division correctly, each term in the numerator needs to be divided by the denominator, 5.

It seems like the two methods yield different answers. The answers look different from each other but they are, in fact, the same. From Method 2, the solution, $x = \frac{b+5y}{5}$, can be rewritten by dividing **each** term in the numerator by the denominator, as follows:

$$x = \frac{b+5y}{5}$$
$$= \frac{b}{5} + \frac{5y}{5}$$
$$= \frac{b}{5} + y$$

giving us the same answer as Method 1.

1.7 Formulas

Exercise 5 Class Example
Solve $3(a+b) = 5$ for a.

Exercise 6 You Try
Solve $2(m-n) = 6p$ for m.

Fractions in Formulas

Formulas often have fractions in them and can be solved in much the same way we solve any equations with fractions. First, identify the lowest common denominator (LCD) and then multiply each term by the LCD to clear the fractions. We can then solve the subsequent equation like any general equation.

Example 5 Solve $\dfrac{a}{3} + \dfrac{b}{4} = c$ for a.

Solution.

We will use the method where we clear the fraction.

$$\frac{a}{3} + \frac{b}{4} = c \qquad \text{Find the LCD}$$

$$(12)\frac{a}{3} + (12)\frac{b}{4} = (12)c \qquad \text{Multiply each term by the LCD} = 12$$

$$4a + 3b = 12c$$

$$4a + 3b - 3b = 12c - 3b \qquad \text{Subtract 3b from each side}$$

$$\frac{4a}{4} = \frac{12c - 3b}{4} \qquad \text{Divide each side by 4}$$

$$x = \frac{12c - 3b}{4} \qquad \text{Our Solution}$$

Note. The solution can be rewritten by dividing **each** term in the numerator by the denominator as follows:

$$a = \frac{12c - 3b}{4}$$

$$= \frac{12c}{4} - \frac{3b}{4}$$

$$= 3c - \frac{3b}{4}$$

It is incorrect to simplify only one term in the numerator. That is, it is incorrect to divide 12c by 4 but not 3b to get $a = 3c - 3b$.

Depending on the context of the problem, we may find a formula that uses the same letter, one capital and one lowercase. These represent different values and we must be careful not to combine a capital variable with a lower case variable.

1.7 Formulas

Exercise 7 Class Example
Solve each formula for the indicated variable.

a) $V = \dfrac{1}{3}bh$ for b

b) $E = \dfrac{mv^2}{2}$ for m

c) $\dfrac{m}{2} - \dfrac{n}{5} = 4$ for n

d) $A = \dfrac{1}{2}h(a+b)$ for b

Exercise 8 You Try
Solve each formula for the indicated variable.

a) $A = \dfrac{1}{2}bh$ for b

b) $P = \dfrac{U-L}{6}$ for L

c) $F = \dfrac{9}{5}C + 32$ for F

d) $\dfrac{x}{3} + \dfrac{y}{6} = z$ for y

1.7: Exercises

Solve each of the following equations for the indicated variable.

1. $P = a + b + c$ for a
2. $I = prt$ for t
3. $S = L + 2B$ for L
4. $E = mc^2$ for m
5. $P = m(n - c)$ for m
6. $V = lwh$ for w
7. $V = \dfrac{\pi D n}{12}$ for D
8. $x + 5y = 3$ for x
9. $ax + b = c$ for x
10. $at - bw = c$ for t
11. $V = \dfrac{1}{3}\pi r^2 h$ for h
12. $A = \dfrac{1}{2} h(a + b)$ for a
13. $5a - 7b = 4$ for a
14. $q = 6(L - p)$ for L
15. $3x + 2y = 7$ for y
16. $C = \dfrac{5}{9}(F - 32)$ for F
17. $A = p + prt$ for r
18. $h = vt - 16t^2$ for v
19. $S = \pi r h + \pi r^2$ for h

Additional Problems

20. The angles x, y, z of a triangle add up to $180°$, that is $x + y + z = 180°$.

    (a) Solve for y.
    (b) Suppose angle $x = 34.6°$ and angle $z = 57.2°$. Find the measure of angle y.

21. The circumference C of a circle with radius r is $C = 2\pi r$.

    (a) Solve for r.
    (b) If a circle has circumference 32 cm, find its radius. (Use $\pi \approx 3.14$.)

22. To find the average A of three tests x, y, z, assuming all tests are weighted equally, we use the formula $A = \dfrac{x + y + z}{3}$.

    (a) Solve for z.
    (b) In order to pass your math class, you have to average 75. Your first two test scores are 78 and 66. What do you need to score on the third test in order to pass the class?

Rescue Roody!

23. Roody is solving for the height h of a pyramid given the volume $V = \frac{1}{3}bh$, where b is the length of the pyramid's base. This is what Roody did, but he was told his answer was not simplified. Roody is confused. Help Roody.

$$A = \frac{1}{3}bh$$
$$\frac{A}{\frac{1}{3}b} = \frac{\frac{1}{3}bh}{\frac{1}{3}b}$$
$$\frac{A}{\frac{1}{3}b} = h$$

24. One of Roody's quiz problems is to use the formula $p = at - b$ to solve for t. Roody is puzzled as to why he received no credit even through he showed his work. This is his work. Help Roody understand his mistake.

$$p = at - b$$
$$\frac{p}{a} = \frac{at}{a} - b$$
$$\frac{p}{a} = t - b$$
$$\frac{p}{a} + b = t$$

1.8 Applications of Linear Equations & Inequalities

Objective: To solve application problems by creating and solving linear equations

Problem Solving Strategies and Tools (PSST)

When looking at an application problem (or story problem), it is often helpful to read the entire problem first and then read it again more slowly to organize your thoughts. The following steps may help in solving an application problem.

A) Identify the unknown quantity and select a variable to represent it.

B) Write an equation or inequality that models the relationship between the known and unknown quantities.

C) Solve the equation or inequality. Check if the solution found is reasonable.

D) Report the solution.

Applications Involving Percents and Fractions

Example 1 A sofa and a love seat together costs $442.50. The love seat is half the cost of the sofa. How much does each cost?

Solution.

Let us go through each step of the Problem Solving Strategies and Tools.

A) *Identify the unknown quantity and select a variable to represent it.*
Let s = cost of sofa
Let $\frac{1}{2}$s = cost of love seat (since love seat is half the cost of the sofa)

B) *Write an equation or inequality that models the relationship between the known and unknown quantities.*

$$\text{cost of sofa} + \text{cost of love seat} = \$442.50$$
$$s + \frac{1}{2}s = \$442.50$$

C) *Solve the equation.*

$s + \frac{1}{2}s = \$442.50$	cost of sofa + cost of love seat = $442.50
$\frac{3}{2}s = \$442.50$	Combine like terms
$\left(\frac{2}{3}\right)\frac{3}{2}s = \$442.50\left(\frac{2}{3}\right)$	Multiply each side by $\frac{2}{3}$, the reciprocal of $\frac{3}{2}$
$s = \$295$	Our Solution (Check reasonableness of solution)

D) *Report the solution.*

The sofa costs \$295 and the love seat costs $\left(\frac{1}{2}\right)(\$295) = \$147.50$

Example 2 Sergeant Piper buys a backpack on sale for \$54.85. The sale price is 31% off the original price. What is the original price?

Solution.

Let us go through each step of the Problem Solving Strategies and Tools.

A) *Identify the unknown quantity and select a variable to represent it.*
Let b = original price of backpack
0.31b = discount price of backpack

B) *Write an equation or inequality that models the relationship between the known and unknown quantities.*

$$\text{Original Price - Discount} = \text{Sale Price}$$
$$b - 0.31b = \$54.85$$

C) *Solve the equation.*

$$\begin{aligned} b - 0.31b &= \$54.85 & &\text{Original Price - Discount = Sale Price} \\ 0.69b &= \$54.85 & &\text{Combine like terms} \\ \frac{0.69b}{0.69} &= \frac{\$54.85}{0.69} & &\text{Divide each side by 0.69} \\ b &= \$79.49 & &\text{Our Solution (Check reasonableness of solution)} \end{aligned}$$

D) *Report the solution.*
Backpack's original price is \$79.49.

1.8 Applications of Linear Equations & Inequalities

Exercise 1 Class Example

After a 57% price reduction, a pair of boots is on sale for $39.90. What is the original price of the boots?

Exercise 2 Class Example

I just realized that my gas gauge says I have $\frac{1}{10}$ of a tank of gas left in my car. I stop at a gas station but can only fill 5 gallons of gas because I did not have enough cash. At this point, my gas gauge shows that my gas tank is halfway full. How big is my gas tank?

Exercise 3 You Try
An online store sells a pair of skis for $299. This is 37% off the original price. What is the original price of the pair of skis?

Exercise 4 You Try
The Washington Huskies have been playing football at the Husky Stadium since 1920. Its seating capacity has been increased by $133\frac{1}{3}\%$ to the present capacity of 70,000. What was the seating capacity of Husky Stadium in 1920?

1.8 Applications of Linear Equations & Inequalities

Geometry

Example 3 The perimeter of a rectangle is 44 cm. The length is 5 cm less than twice the width. Find the dimensions of the rectangle.

Solution.

Let us go through each step of the Problem Solving Strategies and Tools.

A) *Identify the unknown quantity and select a variable to represent it.*
   Let w = width of rectangle
   $2w - 5$ = length of rectangle

B) *Write an equation or inequality that models the relationship between the known and unknown quantities.*
   To find the perimeter of any polygon, add up all its sides.

$$\text{width} + \text{width} + \text{length} + \text{length} = 44$$
$$2 \cdot (\text{width}) + 2 \cdot (\text{length}) = 44$$
$$2 \cdot w + 2 \cdot (2w - 5) = 44$$

C) *Solve the equation.*

$$\begin{aligned}
2w + 2(2w - 5) &= 44 & \\
2w + 4w - 10 &= 44 & \text{Distribute} \\
6w - 10 &= 44 & \text{Combine like terms} \\
6w - 10 + 10 &= 44 + 10 & \text{Add 10 to each side} \\
6w &= 54 & \text{Combine like terms} \\
\frac{6w}{6} &= \frac{54}{6} & \text{Divide each side by 6} \\
w &= 9 & \text{Our Solution (Check reasonableness of solution)}
\end{aligned}$$

D) *Report the solution.*
   The width of rectangle is 9 cm.
   The length of rectangle is 2(9)-5 = 13 cm

Exercise 5 Class Example
The state of Wyoming is roughly in the shape of a rectangle with perimeter 1280 miles. Its width is 80 miles shorter than its length. Find the length and width of Wyoming.

Exercise 6 Class Example
The angles of a triangle add up to 180°. The second angle is twice the first while the third angle is three times the first. Find the measurement of each angle.

Exercise 7 You Try

An isosceles triangle has 2 equal sides. The third side is 10 meters longer than one of the equal sides. The perimeter is 85 meters. Find the measurement of each side.

Exercise 8 You Try

The angles of a triangle add up to 180°. The second angle is 15° less than the first while the third angle is 15° more than the first. Find the measurement of each angle.

Business Applications

In business, to *break even* means that the profit is zero. This will happen when the *cost(s)* incurred to produce the items equals what the business *takes in*, which is called the *revenue*.

Exercise 9 Class Example

Marylou is selling cupcakes for 50¢. the cost to make one cupcake is 22¢. In order to make the cupcakes attractive, Marylou bought 3 dozen special polka dot baking cups that cost her $2.99. How many cupcakes will Marylou need to sell in order to break even?

Exercise 10 You Try

A manufacturer makes specialty pens. It costs 20¢ to produce a pen and their daily fixed cost is $275. The pens are sold for 75¢. How many pens do they need to sell each day to break even?

1.8 Applications of Linear Equations & Inequalities

Inequalities

Example 4 Elmer is a student in Math 089 and needs to spend at least 2 hours each week working on homework or getting help from a tutor at the Math Center. If he spends 35 minutes on Monday, 1 hour and 10 minutes on Thursday, how much more time does Elmer need to spend that week to meet the minimum requirement?

Solution.

Let us go through each step of the Problem Solving Strategies and Tools.

A) *Identify the unknown quantity and select a variable to represent it.*
   Let m = number of minutes needed to meet minimum requirement

B) *Write an equation or inequality that models the relationship between the known and unknown quantities.*
   Time spent on Monday + Time spent on Thursday + additional time needed $\geq$ Minimum time requirement

C) *Solve the equation.*
   Before we begin solving, we need to make sure that the units are all the same. If not, conversion of units is necessary. We know that 1 hour = 60 minutes.
   Minimum time required: 2 hours = 2 hours $\cdot \frac{60 \text{ minutes}}{1 \text{ hour}}$ = 120 minutes
   Time spent on Thursday: 1 hour 10 minutes = 60 minutes + 10 minutes = 70 mintues

   $$\begin{aligned} 35 + 70 + m &\geq 120 & \text{Set up inequality} \\ 105 + m &\geq 120 & \text{Combine like terms} \\ 105 + m - 105 &\geq 120 - 105 & \text{Subtract 105 from each side} \\ m &\geq 15 & \text{Our Solution(Check reasonableness of solution)} \end{aligned}$$

D) *Report the solution.*
   Elmer needs to spend at least 15 more minutes at the Math Center to meet the minimum requirement.

Exercise 11 Class Example
Bob works at a telemarketing firm. On Monday, he made 20 calls. On Tuesday, 28 calls. On Wednesday, 22 calls. On Thursday, 27 calls. What is the minimum number of calls Bob needs to make on Friday to meet the average of at least 25 phone calls per day?

Exercise 12 You Try
Sue needs to get an average of 75 on 4 tests in order to pass her math class. She got a 72 on her first test, 80 on her second test and 68 on her third test. If all the tests are weighted equally, what minimum grade does Sue need to get on the fourth in order to pass her math class?

1.8: Exercises

Solve the following word problems. Be sure to follow the problem solving strategies and tools.

1. A bicycle and a bicycle helmet cost $731.25. How much did each cost, if the bicycle cost 8 times as much as the helmet?

2. A triangle is isosceles if two sides are equal. Find the measurement of each side of an isosceles triangle if the third side is 12 inches shorter than either of the two equal sides and the perimeter of the triangle is 60 inches.

3. A barbecue grill is discounted 25% from its original price and is on sale for $169.95. Find the original price.

4. An elevator in an old apartment building has a maximum capacity of 550 lbs. A worker, weighing 220 lbs, has to deliver water to several apartments and wants to haul as many water bottles as the elevator can hold. If each water bottle weighs 42 lbs, what is the maximum number of bottles he can bring into the elevator each time, assuming no other person enters the elevator?

5. An eight ft board is cut into two pieces. One piece is 2 ft longer than the other. How long are the pieces?

6. Every day, Betty spends $3.85 for an espresso drink. She is thinking about making coffee at home and wonders how quickly she will start to save money. After doing some research, she discovers she can buy an espresso machine for $279 and it would cost 75 cents for each cup. How many cups of coffee will it take before Betty starts saving money by making coffee at home?

7. Seattle wastewater rates for a single family residential customer is $12.27 per 100 cubic feet. According to the Seattle Public Utilities, the typical monthly residential bill is $52.76. At this rate, what is the typical amount of wastewater generated by a single family each month? (Source: seattle.gov)

8. In a room containing 45 students, there were twice as many girls as boys. How many of each were there?

9. A warehouse that stores bags of flour is $\frac{3}{4}$ full. A truck just pulled up and loaded 80 bags of flour for delivery. The warehouse is now $\frac{2}{3}$ full. How many bags of flour can the warehouse store at full capacity?

10. The angles of a triangle sum up to 180°. The second angle of a triangle is 3 times as large as the first angle and the third angle is 30 degrees more than the first angle. Find the measure each angle.

11. The perimeter of a rectangle is 152 meters. The width is 22 meters less than the length. Find the length and width.

12. Lou is selling lemonade for $1.25 a cup. The cost to make one cup of lemonade is 27 cents and each plastic cup is 16 cents. How many cups will he need to sell to make more than $18.75 to watch a movie and buy snacks?

13. Missoula is a big football town with its football team the University of Montana Grizzlies. On one particular game day, 2 hours before game time, the stadium was $\frac{1}{3}$ full. After 30 minutes, an additional 4200 fans entered the stadium. By this time, the stadium was $\frac{1}{2}$ full. What is the stadium's seating capacity?

14. For superbowl 2015, a scalped ticket was being sold for $14,000. This is a 250% mark-up from the original ticket price. What is the original ticket price?

15. May wants to get an average of 95 so she can earn a 4.0 in her math class. Her final grade is the average of 4 tests. Her test scores so far are: 97, 92 and 98. If all the tests are equally weighted, what minimum score does she need in order to earn a 4.0?

16. Jack and Jim began a business with a capital of $7500. If Jack furnished half as much capital as Jim, how much capital did each furnish?

17. Erica is planning a workshop for new business owners. The fee to rent a room is $2075 plus $90 per hour for the catering and support staff. How many hours can she use the room for the workshop and stay within her budget of $2840?

18. A new Honda Civic Hybrid costs $24,500. The car's value depreciates by $2475 every year, on average. How many years can you keep it, if you want to be able to sell it for $2225?

19. Homeowners in a particular city pay property taxes at the rate of $1.48 for every $1000 of assessed value of their home. A homeowner paid $777 in property taxes this year. How much is the homeowner's property worth?

20. A friend works two jobs. In the first job, he tutors 16 hours per week at the LOFT and earns $10.50 per hour. In his second job, he earns $20.00 per house to mow someone's lawn. He asks you (his math friend) to calculate how many lawns he must mow next week to earn at least $353.00 so he can pay his rent?

Puzzle

21. Here is a common mathematical puzzle that is used to stump people. Think of a number, multiply by 2, add 6, divide by 2, and then subtract the number you started with. You should get an answer of 3.

    (a) Try thinking of different numbers to see if the puzzle really works.

    (b) Why does the answer always come out 3? Work it out algebraically.

22. Create your own puzzle and amaze your friends.

Chapter 1 Assessment

Simplify the following expressions.

1. $7 - 4(3-5)^2$

2. $3(2k+1) - (k-5)$

Evaluate each expression using the values given.

3. $pq - (q+5)$; use $p = 2$, $q = -1$

4. $\dfrac{2n}{4m-n}$; use $m = -1$, $n = -4$

Solve the following equations.

5. $6 - w = 12$

6. $3x + 5 = 7x - 5$

7. $3(2y - 1) = 6y$

8. $\dfrac{3h+2}{2} = \dfrac{5}{8} + \dfrac{3}{4}h$

9. $|6p + 3| = 27$

10. $|n + 7| = -1$

Solve each inequality. Write the solution in interval notation and graph the solution on a number line.

11. $-10y < 5$

12. $8 + 5c \leqslant 3c - 10$

Solve each equation for the indicated variable.

13. $3y - 6x = 18$ for y

14. $F = \dfrac{9}{5}C + 32$ for C

Solve the following word problems.

15. A chair is discounted 20% from its original price and is on sale for $675. Find the original cost of the chair.

16. The cost to rent a moving van is $19.95 plus 59¢ per mile driven. How many miles (to the nearest whole mile) can you go if you can afford to pay $50?

2. Graphing Linear Equations

2.1 Introduction to Graphing

Objective: To interpret graphs and work with tables, graphs, and linear equations in two variables

Consider the following data on the median home prices in Seattle.

Year	2011	2012	2013	2014	2015	2016	2017
Price	$353,000	$386,000	$427,000	$459,000	$544,000	$618,000	$718,000

(Source: www.zillow.com)

To get a better sense of how the median home price varies over time, we can build a graph to represent the behavior of the data. In order to do this, we first need to introduce the Rectangular Coordinate System

Rectangular Coordinate System

The **Rectangular Coordinate System**, also known as the **xy-plane**, is divided into four sections by a horizontal number line called the **x-axis** and a vertical number line called the **y-axis.** These four sections are known as the **quadrants**, which are numbered I, II, III and IV counterclockwise. Where the x-axis and the y-axis meet is called the **origin.** This is where $x = 0$ and $y = 0$.
On the x-axis, the positive values are located to the right of the origin and negative values to the left of the origin. On the y-axis, positive values are located above the origin and negative values are located below the origin.

Chapter 2. Graphing Linear Equations

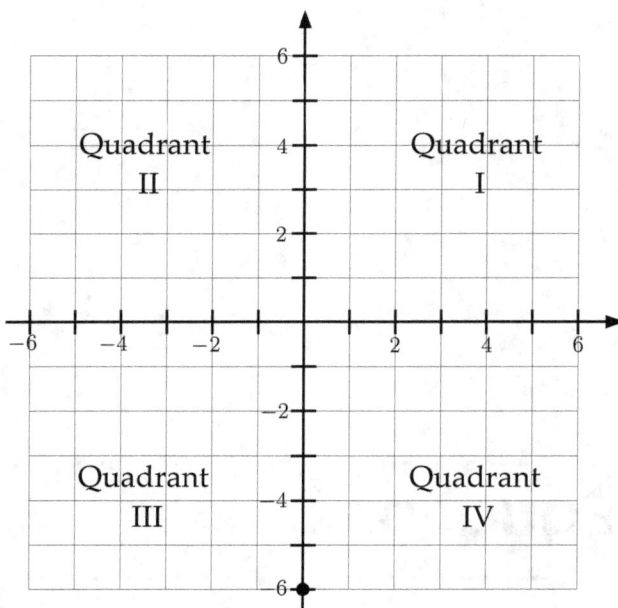

We can describe the location of a point in the plane using the horizontal and vertical signed distances the point is from the origin.

The first number will be called the **x-coordinate**, which is the signed distance the point moves left or right from the origin.

The second number will be the **y-coordinate**, which is the signed distance the point moves up or down from the origin.

The location is given as an **ordered pair**, written as (x, y).

World View Note French mathematician Rene Descartes, was who first proposed to describe each point on the plane by two numbers, giving the points horizontal and vertical location. This has come to be known as the Cartesian coordinates.

Example 1 Find the ordered pair that describes the location of each point.
Solution.

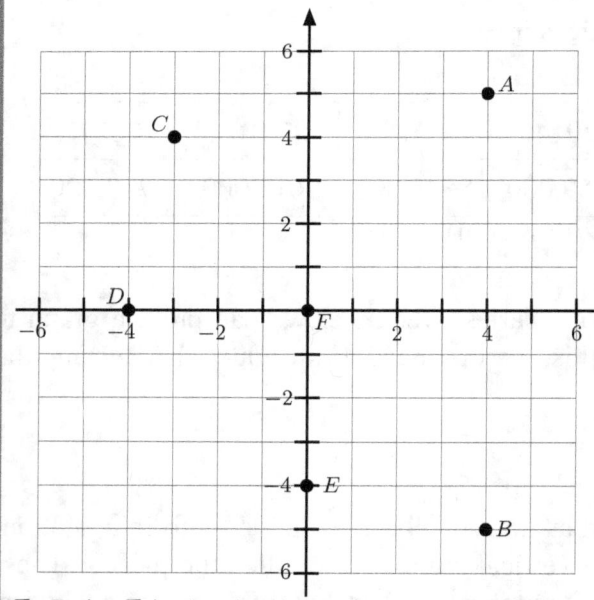

A: Starting from the origin, point A is 4 units to the right and 5 units up. We write this as the ordered pair $(4, 5)$.

B: Likewise, point B is 4 units to the right and 5 units down from the origin. So its ordered pair is $(4, -5)$.

C: From the origin, point C is 3 units to the left and 4 units up. So its ordered pair is $(-3, 4)$.

D: Point D is 4 units directly to the left of the origin. We do not move up or down. Thus, we write $(-4, 0)$ for its ordered pair.

E: Point E is 4 units down from the origin. We do not move left or right. So $(0, -4)$ is its ordered pair.

F: Finally, point F is located at the origin, so its ordered pair is $(0, 0)$.

2.1 Introduction to Graphing

Note. Order matters with ordered pairs! In the previous example, the points $B = (4, -5)$ and $C = (-5, 4)$ have reversed coordinates and so describe different locations. It is always important to remember that the **x-coordinate** is listed **first** and the **y-coordinate** is listed **second.**

Example 2 Find the ordered pair that describes the location of each point.
Solution.

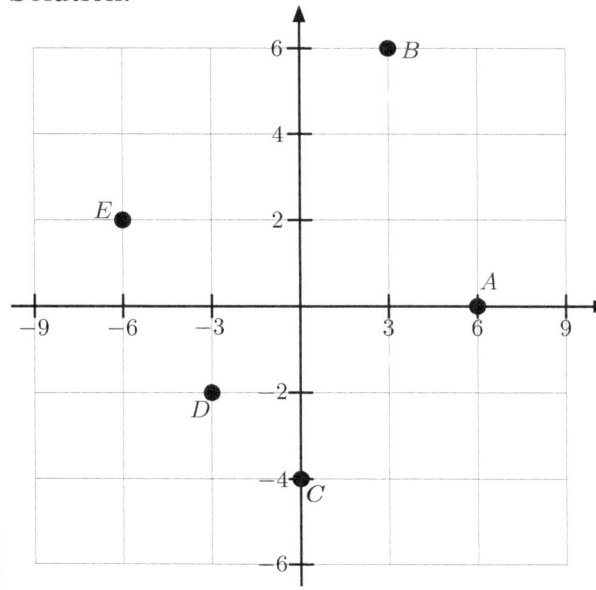

Here, each tick mark on the x-axis is a multiple of 3 whereas each tick mark on the y-axis is a multiple of 2. Because of this, we must take care in labeling points accordingly.
The ordered pairs for the points shown are as follows.

$A = (6, 0)$, $B = (3, 6)$, $C = (0, -4)$, $D = (-3, -2)$, and $E = (-6, 2)$.

Exercise 1 **Class Example**
Find the ordered pair that describes the location of each point. Also, state which quadrant each point lies in or whether the point lies on the x-axis or y-axis. Pay close attention to the scale on each axis.

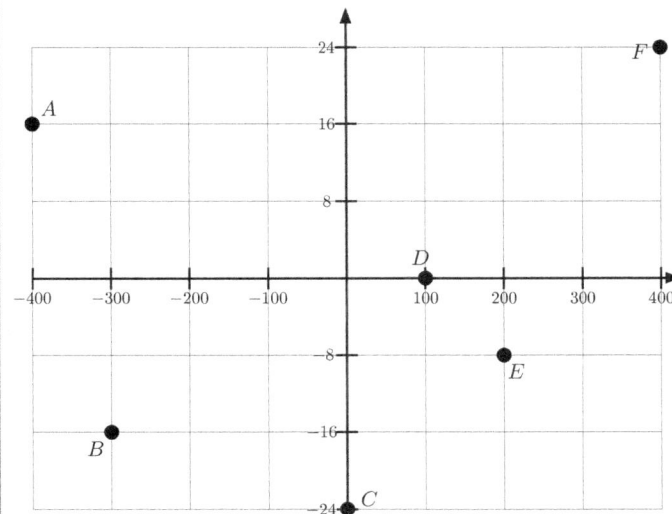

Exercise 2 You Try
Plot the ordered pairs $A = (-4, -2)$, $B = (3, 8)$, $C = (1, -6)$, and $D = (0, 6)$. Be sure to pay attention to the scale on each axis.

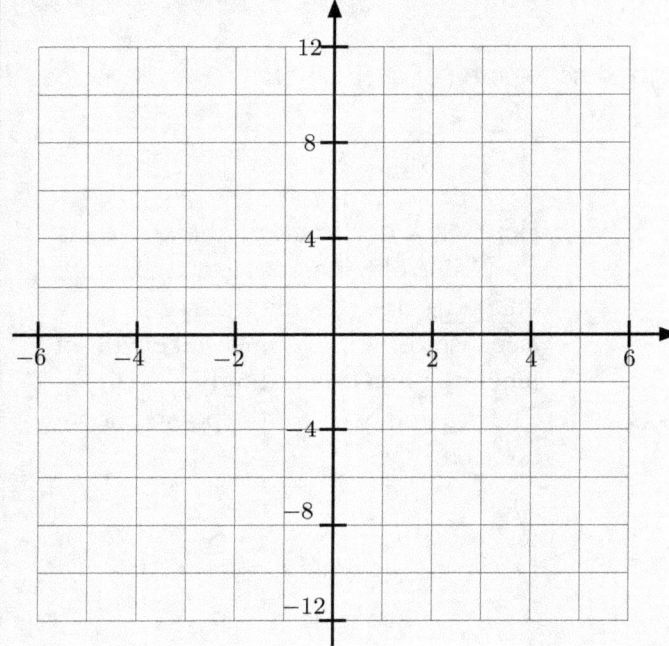

Exercise 3 You Try
Suppose the point (a, b) is located in Quadrant I.

a) In which quadrant is the point $(a, -b)$ located?

b) In which quadrant is the point $(-a, -b)$ located?

2.1 Introduction to Graphing

Exercise 4 **You Try** Find the ordered pair that describes the location of each point. Pay attention to the scale on each axis. You may need to estimate the location of some points.

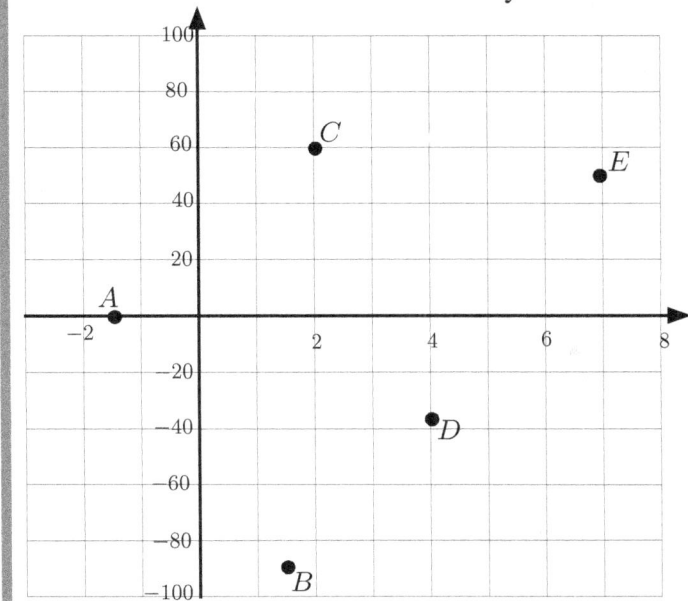

Graph of an Equation in Two Variables

In the previous chapter, we dealt with equations in one variable such as $2x - 3 = 5$. The solution to such an equation is a real number ($x = 4$ for this equation), and we can plot this on the number line. In this chapter, we introduce **equations in two variables**, such as $y = 4x + 7$, $d = 30t$, and $a - b = 5$. To specify a solution to this type of equation, we now need to give two real numbers that make the equation true after substitution. For example, if we substitute $x = 1$ and $y = 11$ into the equation $y = 4x + 7$, we get a true statement since $11 = 4(1) + 7$. Below is a table containing a few other solution pairs to this equation. We can think of each solution pair to the equation, $y = 4x + 7$, as an ordered pair.

x	$y = 4x + 7$	Ordered Pair
0	$y = 4(0) + 7 = 7$	$(0, 7)$
-2	$y = 4(-2) + 7 = -1$	$(-2, -1)$
3	$y = 4(3) + 7 = 19$	$(3, 19)$
$\frac{1}{2}$	$y = 4(\frac{1}{2}) + 7 = 9$	$(\frac{1}{2}, 9)$

In the above example, we say that the ordered pairs $(0, 7)$, $(-2, -1)$, $(3, 19)$, and $(\frac{1}{2}, 9)$ are all solutions to the equation, $y = 4x + 7$.

> A **solution** to an equation in two variables is an ordered pair such that the coordinates of the ordered pair make the equation true, after substitution.

Example 3 Are the following ordered pairs solutions to the equation $3y - 2x = 23$?
a) $(-4, 5)$
b) $\left(6, \dfrac{5}{3}\right)$

Solution.

a) Check if $(x, y) = (-4, 5)$ is a solution to the equation, $3y - 2x = 23$.

$$3y - 2x = 23 \qquad \text{Substitute } x = -4 \text{ and } y = 5$$
$$3(5) - 2(-4) \stackrel{?}{=} 23 \qquad \text{Perform indicated operation}$$
$$15 + 8 = 23 \checkmark \qquad (-4, 5) \text{ is a solution to } 3y - 2x = 23$$

b) Check if $(x, y) = \left(6, \dfrac{5}{3}\right)$ is a solution to the equation, $3y - 2x = 23$.

$$3y - 2x = 23 \qquad \text{Substitute } x = 6 \text{ and } y = \dfrac{5}{3}$$
$$3\left(\dfrac{5}{3}\right) - 2(6) \stackrel{?}{=} 23 \qquad \text{Perform indicated operation}$$
$$5 - 12 \neq 23 \qquad \left(6, \dfrac{5}{3}\right) \text{ is not a solution to } 3y - 2x = 23$$

Exercise 5 Class Example
Are the following ordered pairs solutions to the equation $2x - y = 5$?
a) $(1, 3)$
b) $\left(\dfrac{1}{2}, -4\right)$

2.1 Introduction to Graphing

Exercise 6 You Try

Are the following ordered pairs solutions to the equation $y = -\frac{2}{3}x + 5$?

a) $(-3, 3)$

b) $\left(-\frac{9}{2}, 8\right)$

2.1: Exercises

Find the ordered pair that describes the location of each point.

1.
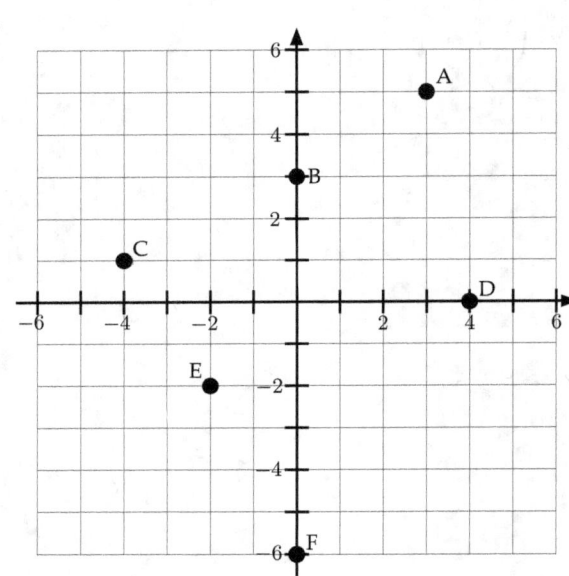

2.
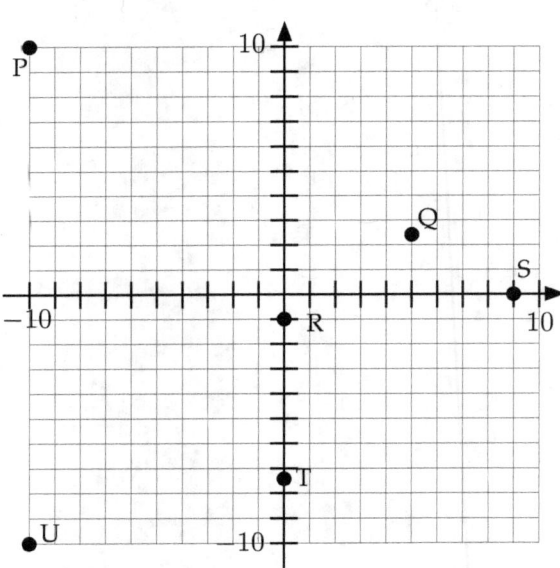

Plot the following ordered pairs and tell in what quadrant or on what axis the point is located.

3. A $(5,3)$

4. B $(-1,3)$

5. C $((4,-2)$

6. D $(0,-7)$

7. E $(-2,-8)$

8. F $(0,3)$

9. G $(-6,0)$

10. H $(1,4)$

11. I $((-5,-2)$

12. J $(9,0)$

2.2 Graphing Linear Equations in Two Variables

Objective: To be able to produce a table of solutions to a linear equation in two variables, graph those points and produce a line graph

A **linear equation in two variables** can be written as $Ax + By = C$, where A, B, and C are any real numbers. As you learned in the previous section, the solutions to this type of equations are ordered pairs, (x, y), whose values make the equation true after substitution. We can visualize the solutions to an equation in two variables by plotting the ordered pairs. For example, consider the equation $y = 2x$ and the given table of solutions. Plotting the ordered pair solutions and connecting the points with a line gives us the **graph** of the equation.

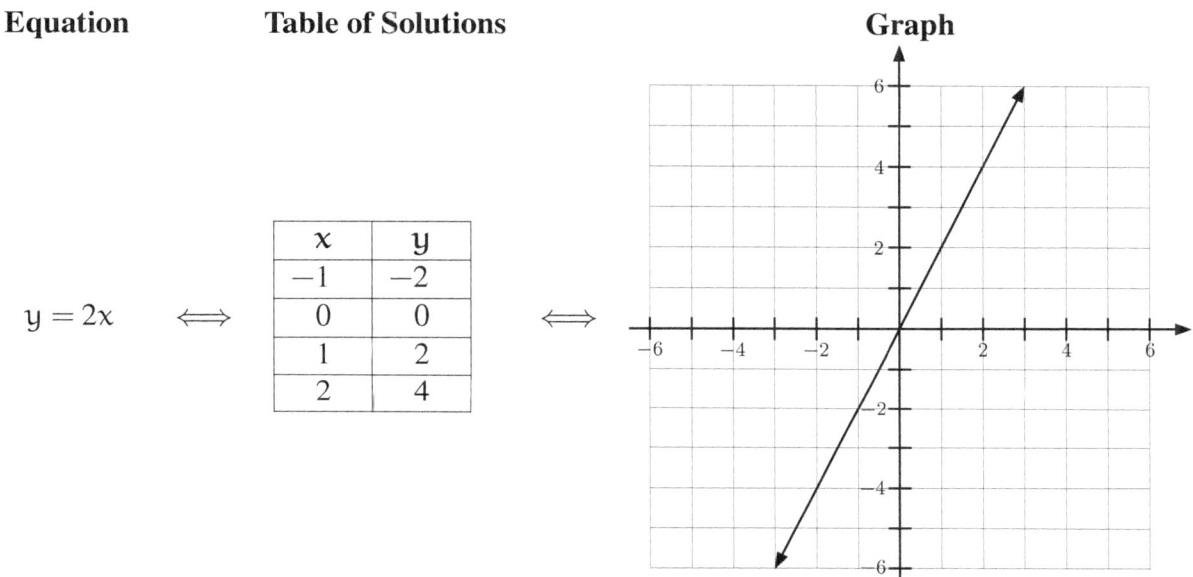

Equation **Table of Solutions** **Graph**

$y = 2x$ $\iff$

x	y
−1	−2
0	0
1	2
2	4

$\iff$

How many solutions are there to a linear equation in two variables? An infinite number of solutions. We can list some of them in a table. However, it also helps to be able to see them on a graph.

In fact, the relationship between two variables can be expressed in any of these three ways:

- as an equation
- as a table of solutions, or
- as a graph.

Ultimately, you will be able to start with any one of these and produce either of the other two.

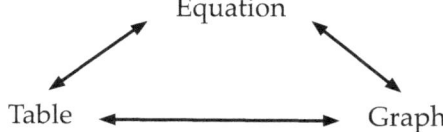

Exercise 1 Class Example

Given the graph of the equation, $y = -x + 4$, produce a table of 3 solutions by reading the points off the graph. Then confirm that each ordered pair solves the given equation.

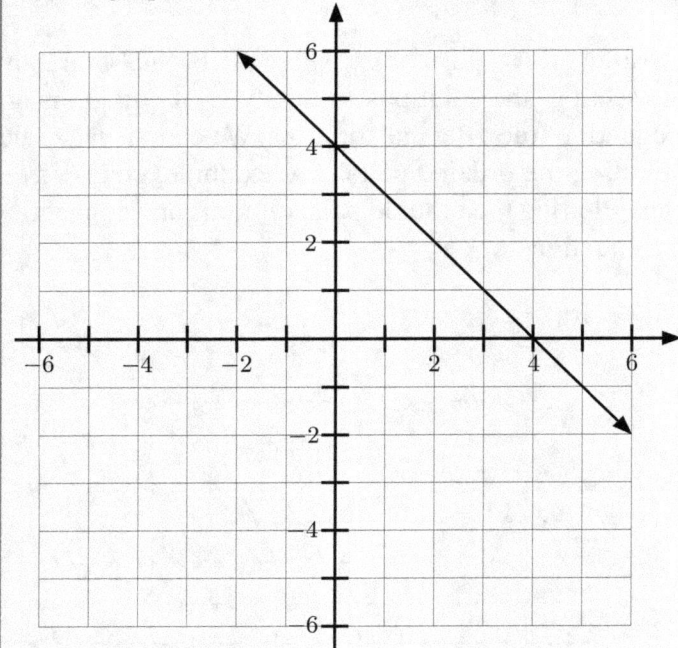

Example 1 Graph the equation, $2x + y = 1$ by first producing a table of at least 3 solutions.

Solution.
We start by producing a table of at least 3 solutions. Two points define a line but the third point ensures that our line is correct. Substitute a value for either the x or the y coordinate and then solve for the other coordinate. It does not matter what value we choose to start with, though some choices are easier to graph.

Let us begin with $x = 1$ and solve for the y.

$2x + y = 1$	Substitute $x = 1$
$2(1) + y = 1$	Perform the indicated operation
$2 + y = 1$	Subtract 2 from each side
$y = -1$	Our y-coordinate

One solution to the equation is $(1, -1)$.

Next, let $y = 5$ and solve for x.

$2x + y = 1$	Substitute $y = 5$
$2x + (5) = 1$	Subtract 5 from each side
$2x = -4$	Divide each side by 2
$x = -2$	Our x-coordinate

2.2 Graphing Linear Equations in Two Variables

Another solution to the equation is $(5, -2)$.

Next, let $y = -2$ and solve for x.

$$\begin{aligned} 2x + y &= 1 & &\text{Substitute } y = -2 \\ 2x + (-2) &= 1 & &\text{Add 2 to each side} \\ 2x &= 3 & &\text{Divide each side by 2} \\ x &= \frac{3}{2} & &\text{Our x-coordinate} \end{aligned}$$

Another solution to the equation is $\left(\frac{3}{2}, -2\right)$. Fractions are not always easy to graph, so we will look for another point.

Let $x = -1$ and solve for y.

$$\begin{aligned} 2x + y &= 1 & &\text{Substitute } x = -1 \\ 2(-1) + y &= 1 & &\text{Perform the indicated operation} \\ -2 + y &= 1 & &\text{Add 2 to each side} \\ y &= 3 & &\text{Our y-coordinate} \end{aligned}$$

Another solution to the equation is $(-1, 3)$.

Lets now put these solutions into a table, and use them to create the graph of the line for this equation.

Table of Solutions

x	y	Ordered Pair
-2	5	$(-2, 5)$
-1	3	$(-1, 3)$
1	-1	$(1, -1)$
$\frac{3}{2}$	-2	$\left(\frac{3}{2}, -2\right)$

Graph

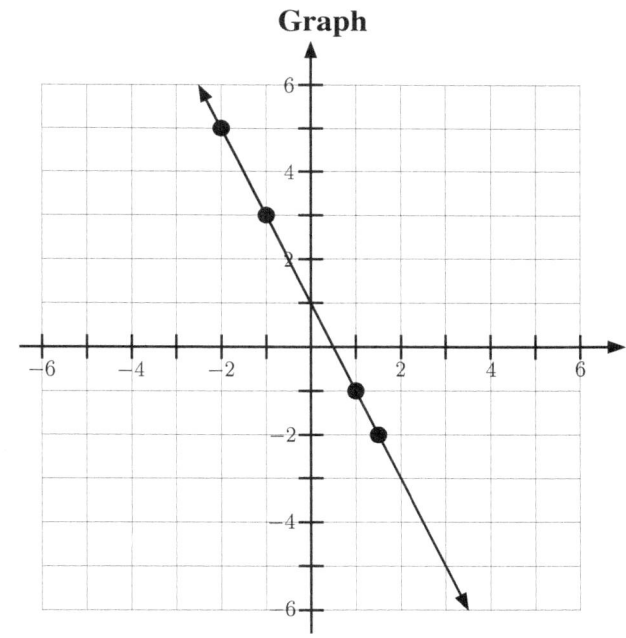

Exercise 2 Class Example
Graph the equation, $y = 3x$ by first producing a table of at least 3 solutions.

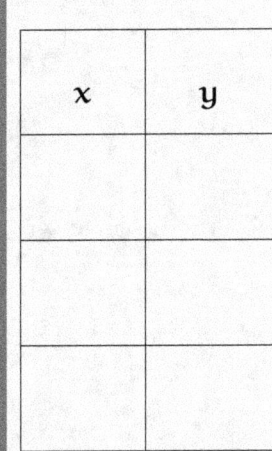
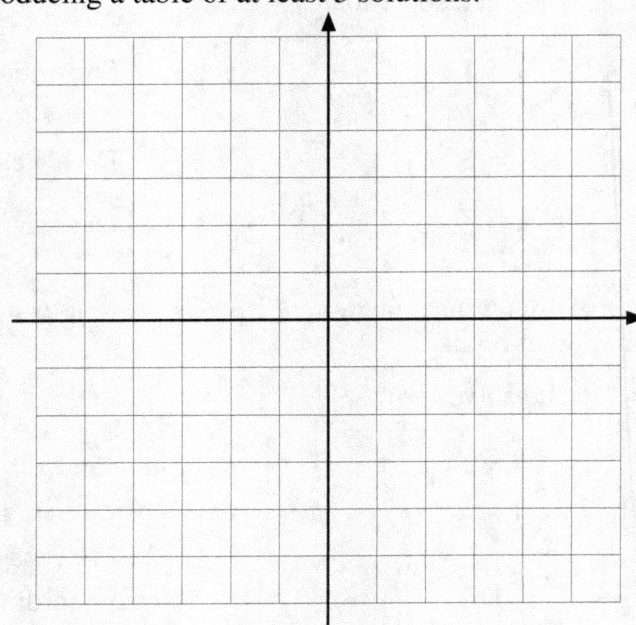

Exercise 3 Class Example
Graph the equation, $x - 2y = 4$ by first producing a table of at least 3 solutions.

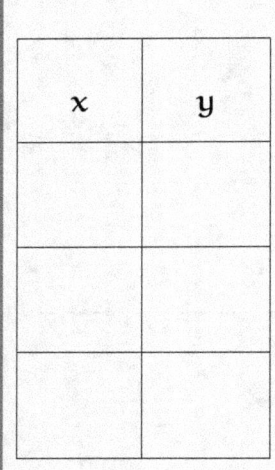
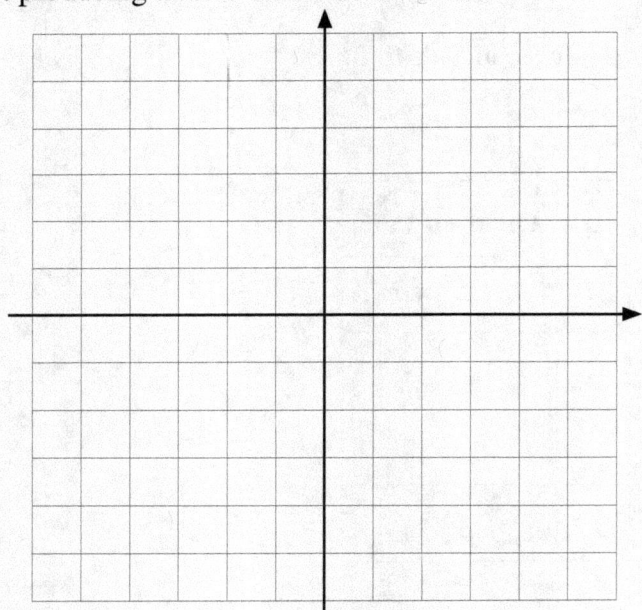

2.2 Graphing Linear Equations in Two Variables

Exercise 4 You Try
Graph the equation, $y = 2x - 3$ by first producing a table of at least 3 solutions.

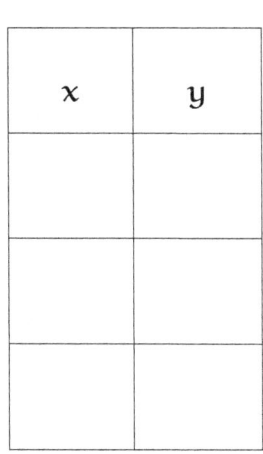
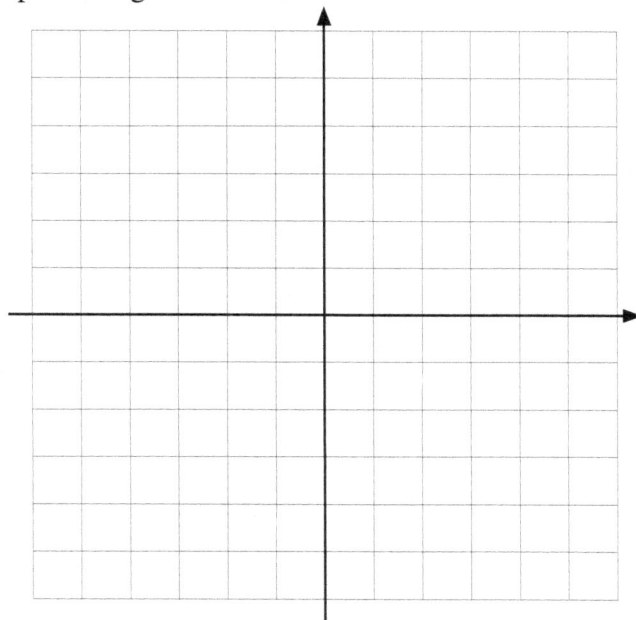

Exercise 5 You Try
Graph the equation, $x - y = 0$ by first producing a table of at least 3 solutions.

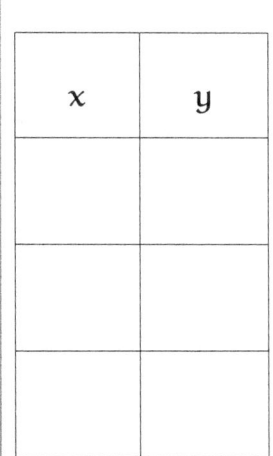
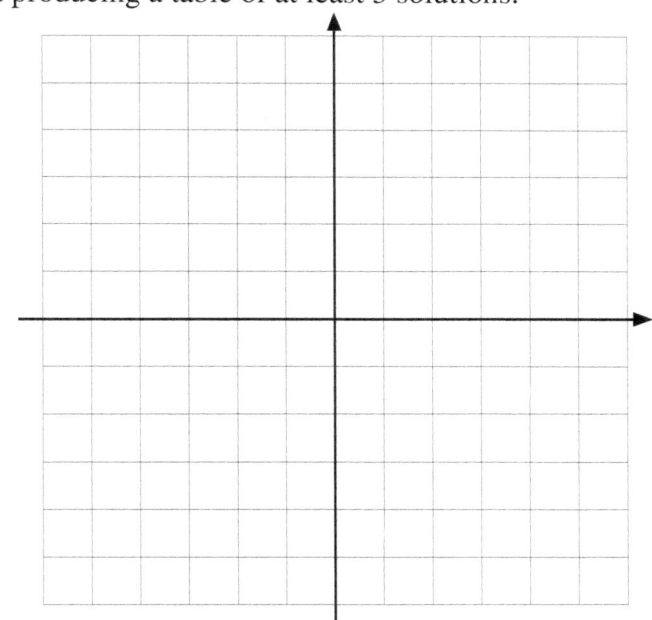

Vertical and Horizontal Lines

Consider the following linear equation, $y = 2$. The first question you might ask is "Where is the other variable?" Remember the form of a linear equation in two variables is $Ax + By = C$. In this case, $A = 0$.

Example 2 Graph $y = 2$.

Solution.

Let us find at least 3 points first by letting $x = -1$. Since there is no x term in the equation, there is no place to substitute $x = -1$. Therefore, $y = 2$.

What about when $x = 0$? Same answer, $y = 2$.

What about when $x = 1$? Same answer, $y = 2$

What about when $x = \frac{3}{2}$? Same answer, $y = 2$.

The pattern we are seeing here is that whatever x-value we choose, the y-value is always 2. Let us put these solutions into a table and use them to sketch the graph.

Table of Solutions

x	y	Ordered Pair
-1	2	$(-1, 2)$
0	2	$(0, 2)$
1	2	$(1, 2)$
$\frac{3}{2}$	2	$(\frac{3}{2}, 2)$

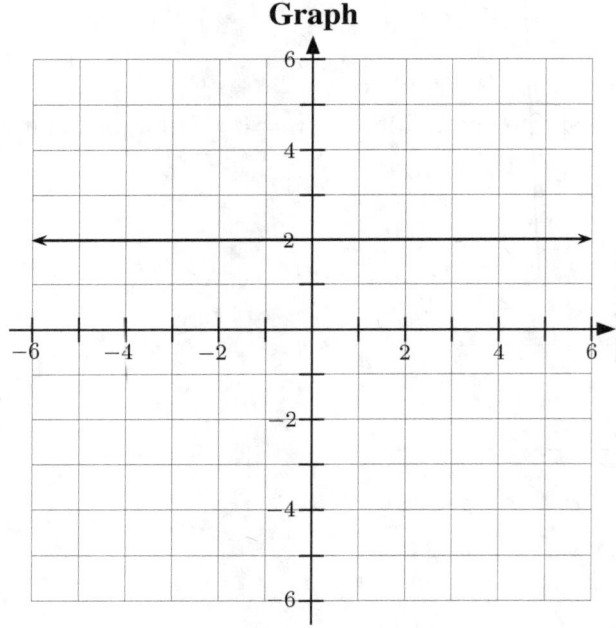

Graph

Example 3 Graph $x = -1$.

Solution.

Let us find at least 3 points first by letting $y = -2$. Since there is no y term in the equation, there is no place to substitute $y = -2$. Therefore, $x = -1$.

What about when $y = 0$? Same answer, $x = -1$.

What about when $y = 0.5$? Same answer, $x = -1$

What about when $y = 3$? Same answer, $x = -1$.

The pattern we are seeing here is that whatever x-value we choose, the y-value is always 2. Let us put these solutions into a table and use them to sketch the line.

Table of Solutions

x	y	Ordered Pair
−1	−1	(−1, −1)
0	−1	(0, −1)
0.5	−1	(0.5, −1)
3	−1	(3, −1)

Graph

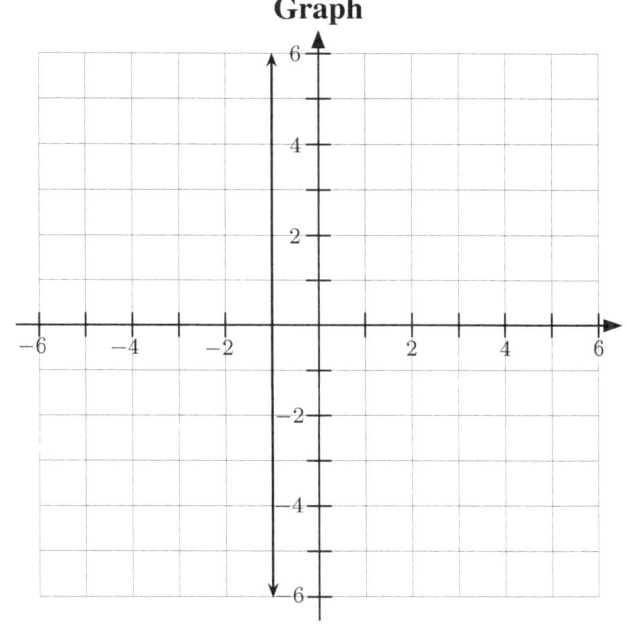

Lets summarize what we have seen.

- A line where all the y-coordinates are equal is a **horizontal line** and its equation takes the form, $y =$ a real number.
- A line where all the x-coordinates are equal is a **vertical line** and its equation takes the form, $x =$ a real number.

Chapter 2. Graphing Linear Equations

Exercise 6 Class Example
Graph the following lines by first producing a table of at least 3 solutions.

a) $x = 3$

x	y

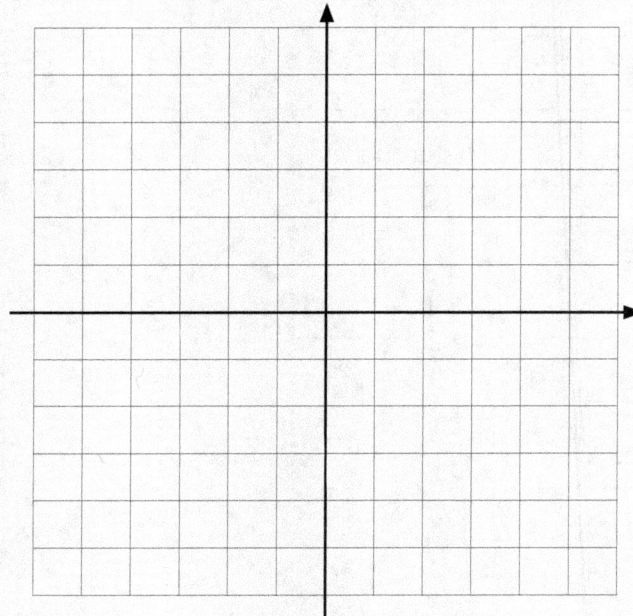

b) $y = -4$

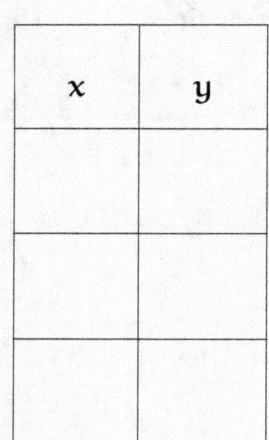

x	y

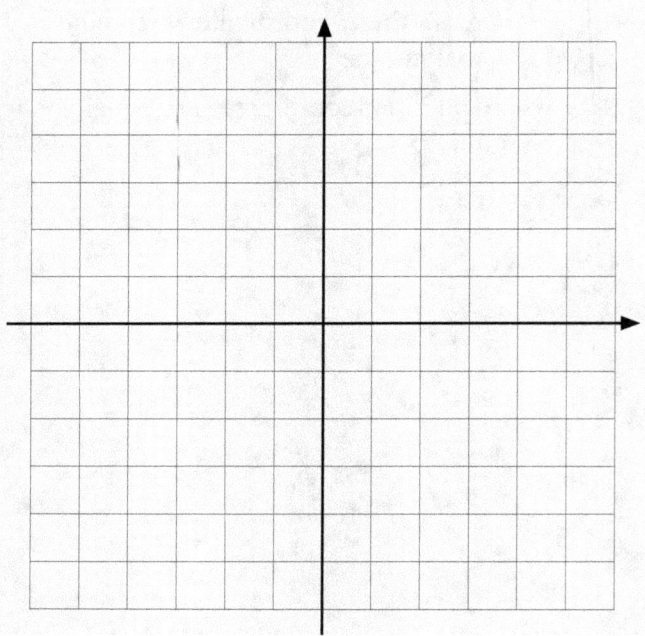

2.2 Graphing Linear Equations in Two Variables

Exercise 7 Class Example
Graph the following lines by first producing a table of at least 3 solutions.

a) $y = 1$

x	y

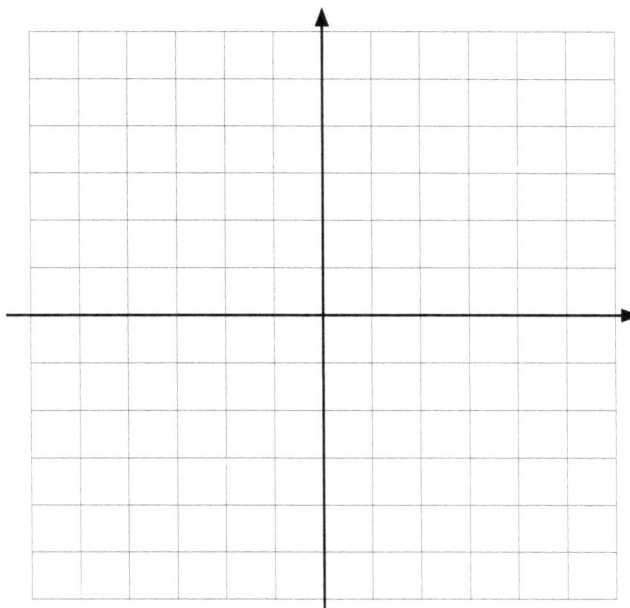

b) $x = -2$

x	y

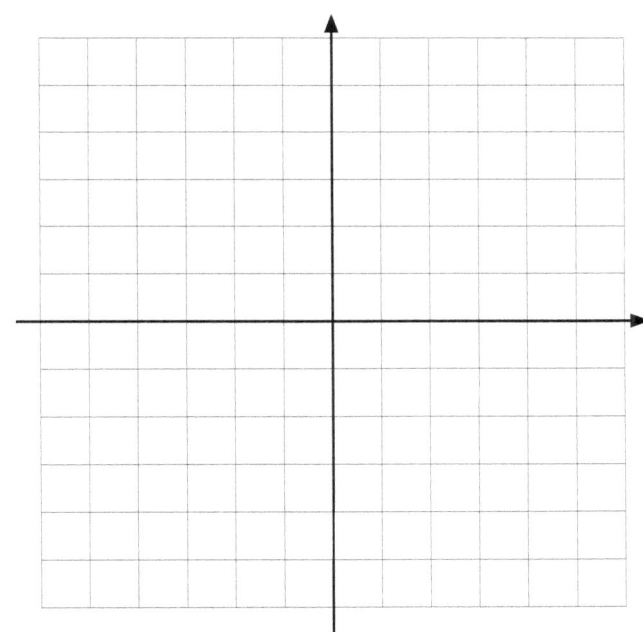

c) $x = 0$

x	y

d) $y = 0$

x	y

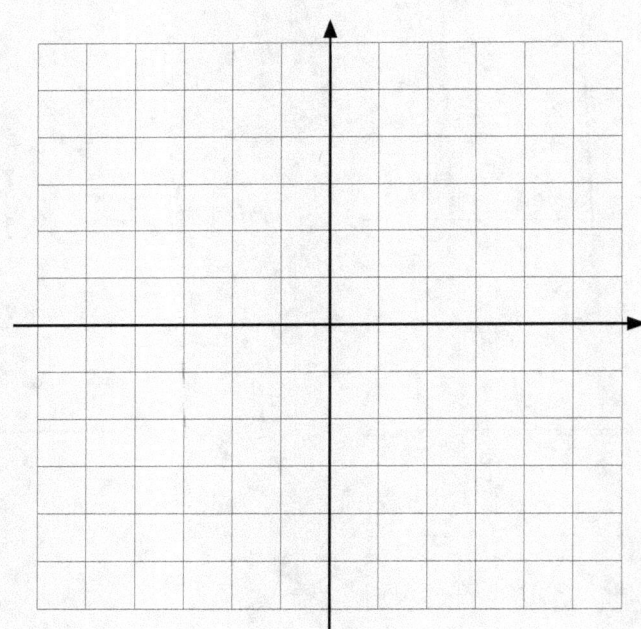

2.2 Graphing Linear Equations in Two Variables

2.2: Exercises

Complete the table using the given equation, and use the table to graph the equation.

1. $y = 2x$

x	y
−1	
0	
1	

2. $y = -x + 2$

x	y
−3	
	3
2	

3. $x - 3y = -5$

x	y
−2	
1	
	3

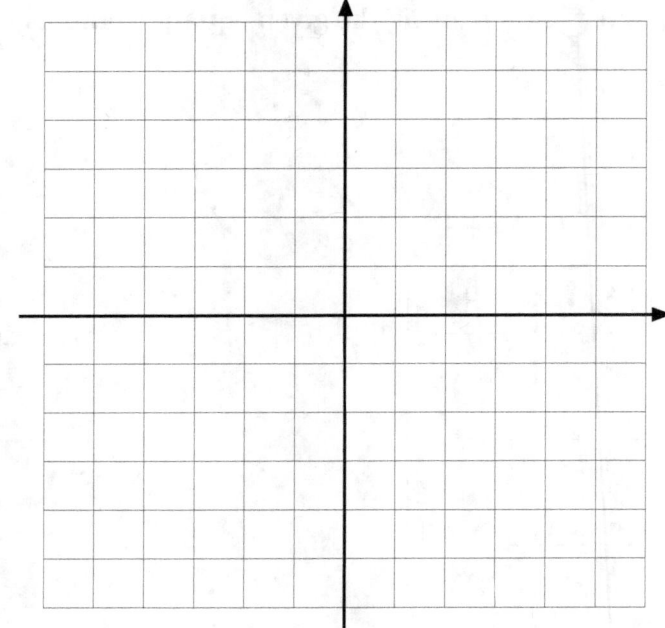

4. $y = 4$

x	y
−2	
−1	
3	

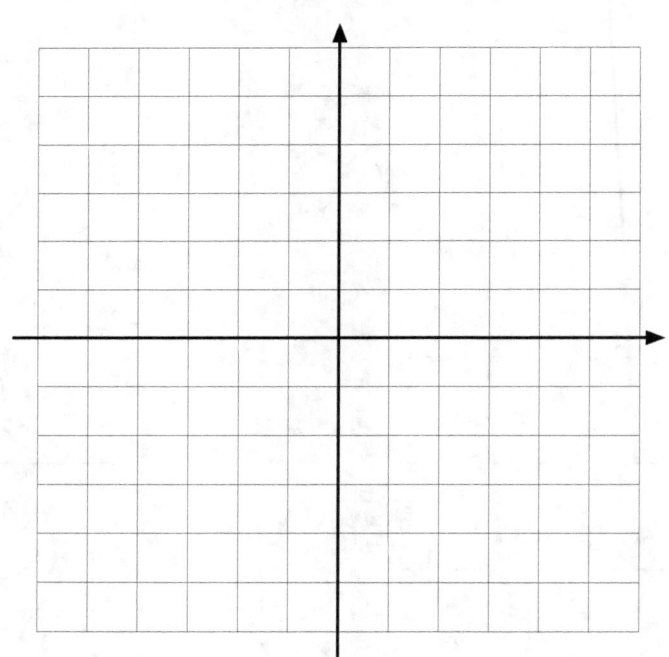

2.2 Graphing Linear Equations in Two Variables

5. $x = -3$

x	y
	-4
	1
	2

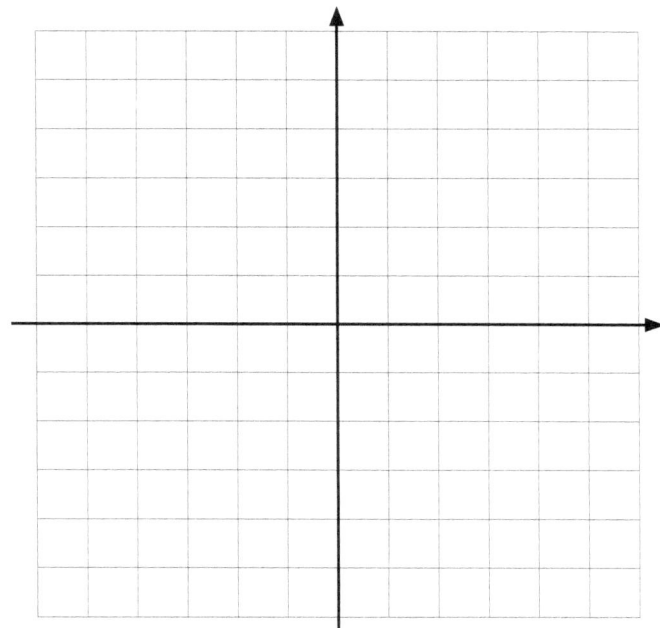

6. $y = -2x + 1$

x	y

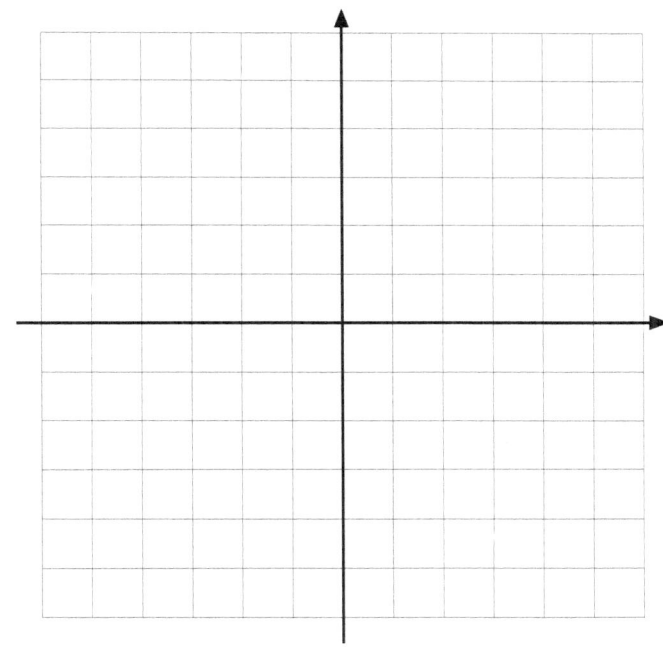

2.3 Intercepts

Objective: To find the x-intercept and y-intercept and use them to produce a line graph

The **y-intercept** of a line is the point where the line crosses the y-axis. The x-coordinate at that point is 0. The **x-intercept** of a line is the point where the line crosses the x-axis. The y-coordinate at that point is 0.

Example 1 Determine the x-intercept and y-intercept of the line. Write your answers as ordered pairs.

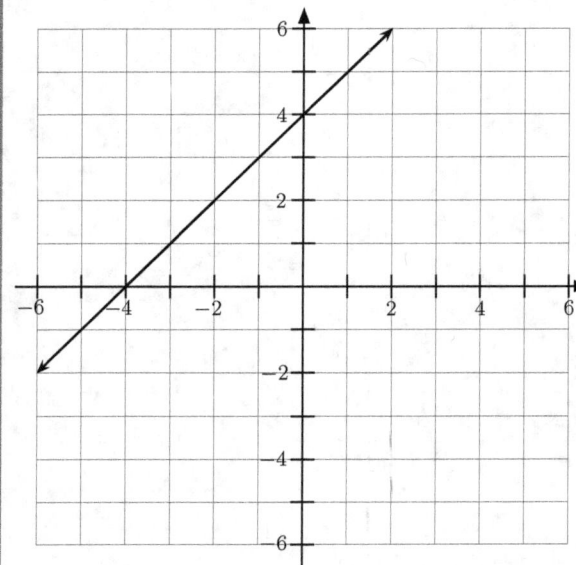

Solution.
x-intercept: The line intersects the x-axis at the point $(-4, 0)$. This is the x-intercept of the line. Notice that the y-coordinate of the point is 0.

y-intercept: The line intersects the y-axis at the point $(0, 4)$. This is the y-intercept of the line. Notice that the x-coordinate of the point is 0.

Exercise 1 Class Example
Determine the x-intercept and y-intercept of the line. Write your answers ordered pairs.

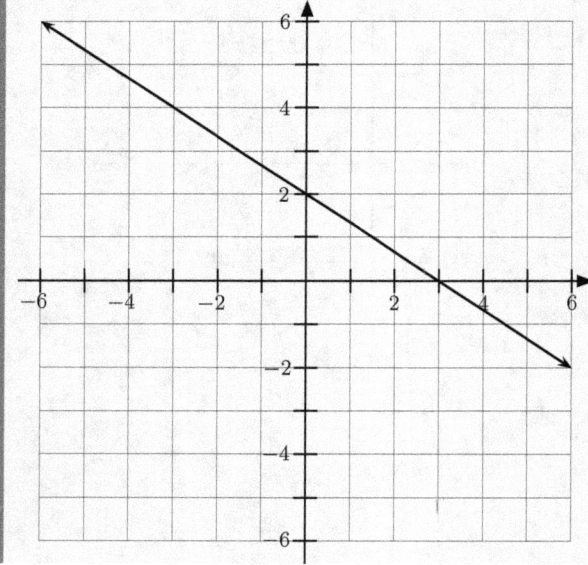

2.3 Intercepts

Exercise 2 You Try

Determine the x-intercept and y-intercept of the line. Write your answers as ordered pairs.

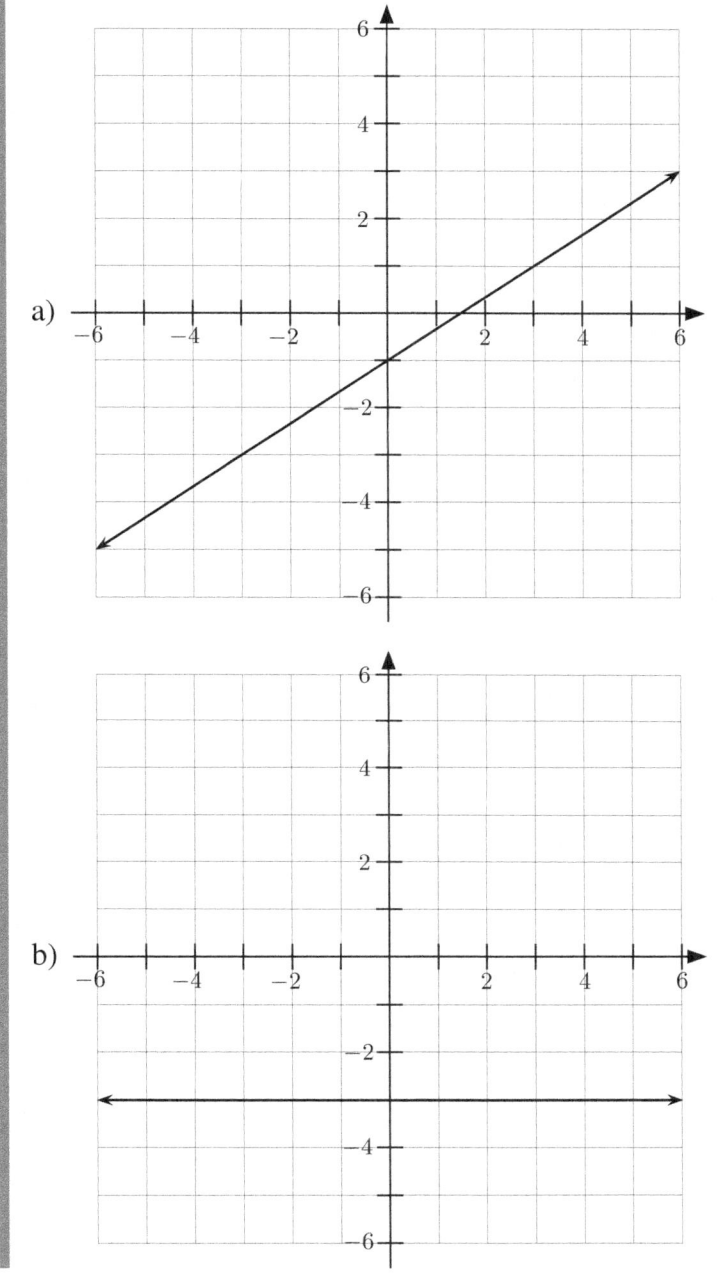

a)

b)

Given an equation, how do we find the intercepts?

> To find the x-intercept, let $y = 0$ and solve for x.
> To find the y-intercept, let $x = 0$ and solve for y.

Example 2 Determine the x-intercept and y-intercept of the line, $x + 2y = 4$. Write your answers as ordered pairs.

Solution.
To find the x-intercept, let $y = 0$ and solve for x.

$x + 2y = 4$	Substitute $y = 0$
$x + 2(0) = 4$	Perform the indicated operation
$x = 4$	The x-intercept is $(4, 0)$

To find the y-intercept, let $x = 0$ and solve for y.

$x + 2y = 4$	Substitute $x = 0$
$(0) + 2y = 4$	Perform the indicated operation
$2y = 4$	Divide each side by 2
$y = 2$	The y-intercept is $(0, 2)$

Exercise 3 Class Example
Determine the x-intercept and y-intercept of the linear equation. Write your answer as an ordered pair.

a) $y = -3x$

b) $y = \frac{1}{2}x + 3$

Exercise 4 You Try
Determine the x-intercept and y-intercept of the linear equation. Write your answer as an ordered pair.

a) $y = 2x - 3$

b) $x - y = 0$

Graphing Lines Using Intercepts

We can use the x and y intercepts to graph a line. If the x-intercept is different from the y-intercept, these two points will define the line. To ensure you have the correct line, find a third point. If the x-intercept is the same as the y-intercept, which happens when both intercepts are at the origin,

2.3 Intercepts   119

you will need to find one additional point to be able to sketch the line and another point to ensure you have the correct line.

Example 3 Find the x and y-intercepts. Then use those points to graph the line, $3x - y = 6$.

Solution.
Let us find the x-intercept by setting $y = 0$ and solving for x.

$3x - y = 6$	Substitute $y = 0$
$3x - (0) = 6$	Perform indicated operation
$3x = 6$	Divide each side by 3
$x = 2$	The x-intercept is $(2, 0)$

Let us find the y-intercept by setting $x = 0$ and solving for y.

$3x - y = 6$	Substitute $x = 0$
$3(0) - y = 6$	Perform indicated operation
$-y = 6$	Multiply each side by -1
$y = -6$	The y-intercept is $(0, -6)$

Find a third point to ensure we have the correct line. Let $x = 1$ and solve for y.

$3x - y = 6$	Substitute $x = 1$
$3(1) - y = 6$	Perform indicated operation
$3 - y = 6$	Subtract 3 from each side
$-y = 3$	Multiply each side by -1
$y = -3$	A third point is $(1, -3)$

Now that we have at least 3 points, we can graph the line that represents the equation, $3x - y = 6$.

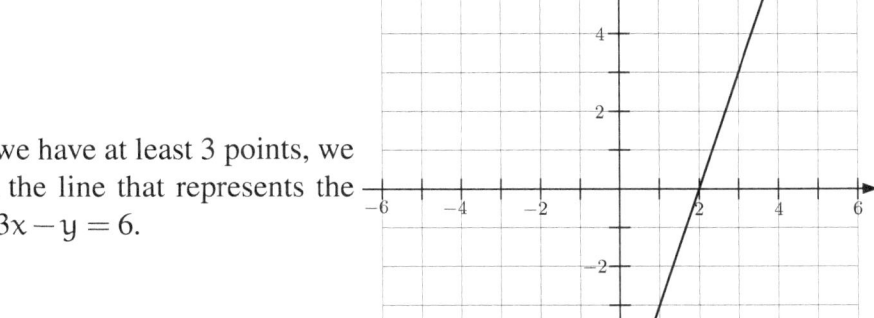

Exercise 5 Class Example
For each equation, find the x-intercept and y-intercept. Use those points to graph the line.

a) $y = \dfrac{1}{2}x + 1$

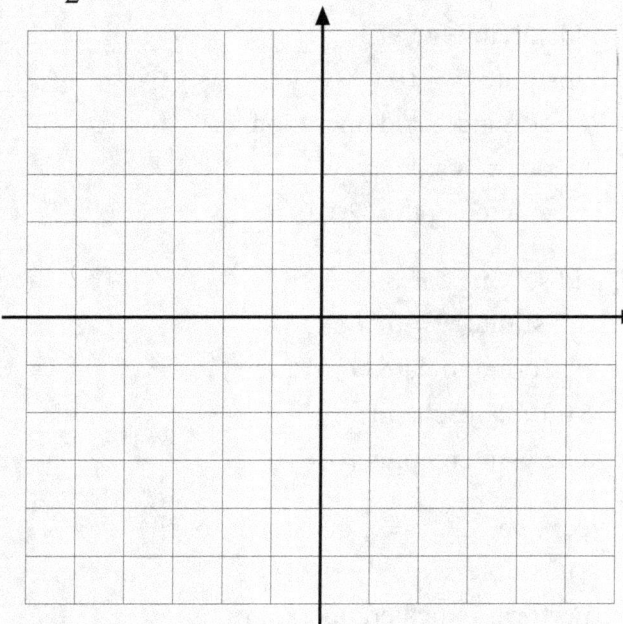

b) $3x + y = 6$

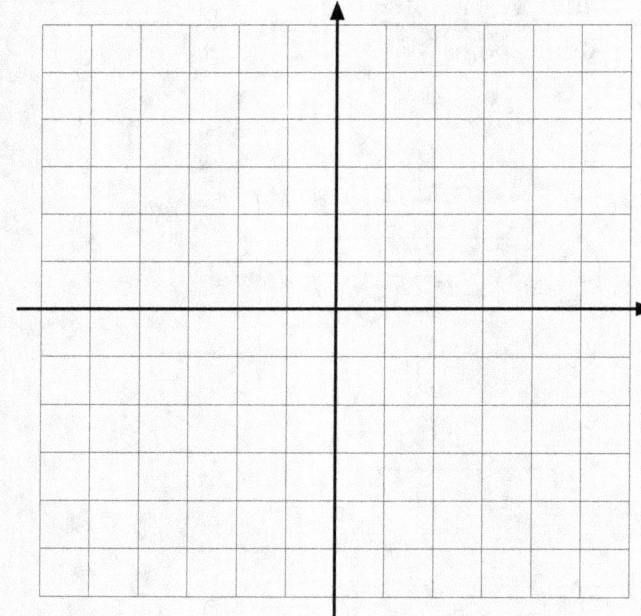

2.3 Intercepts

Exercise 6 You Try

For each equation, find the x-intercept and y-intercept. Use those points to graph the line.

a) $x + y = 3$

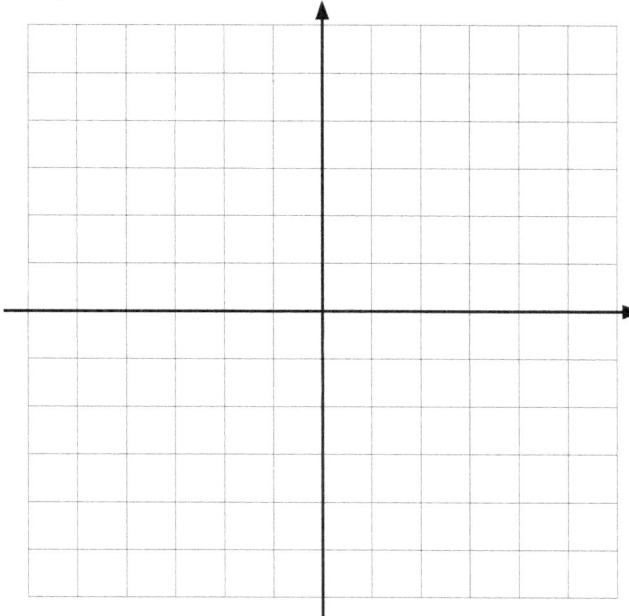

b) $y = 5x - 5$

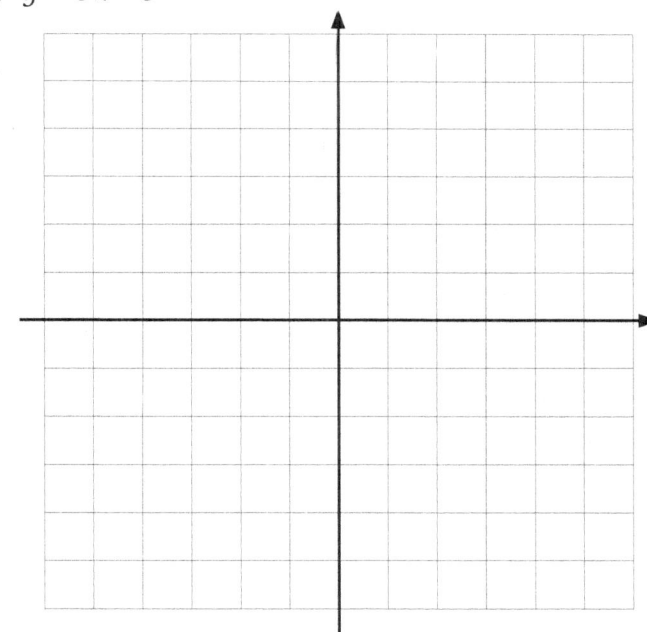

2.3: Exercises

For each graph, identify the x- and y-intercepts, if they exits. Write your answers as ordered pairs.

1.
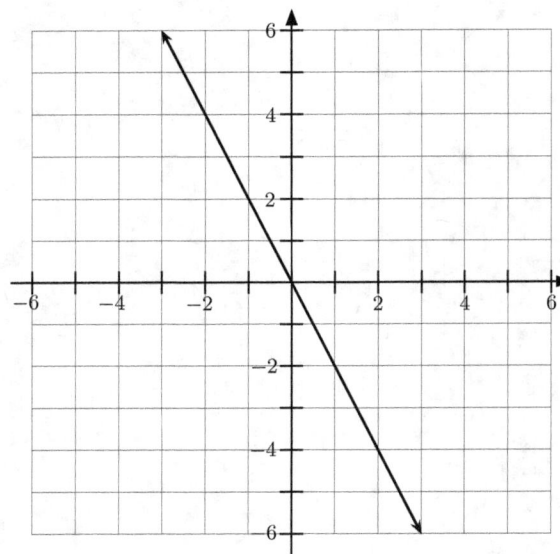

3.
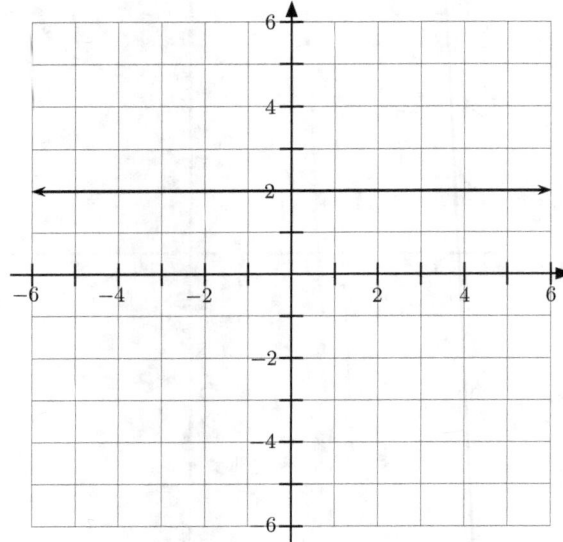

2.
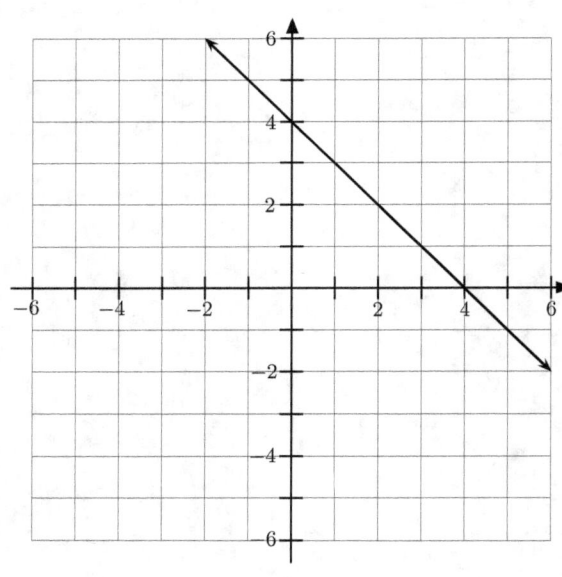

4.
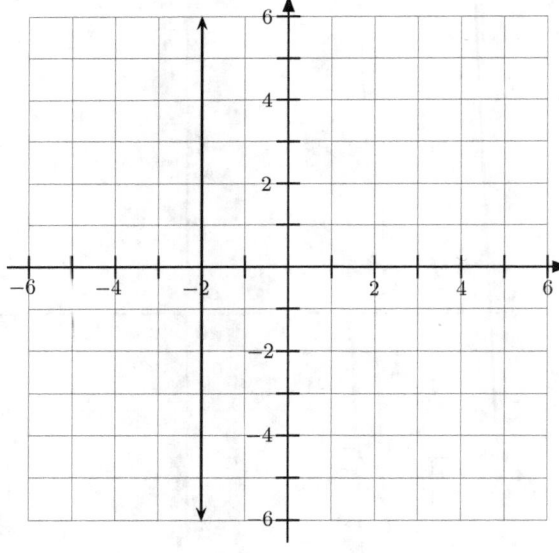

2.3 Intercepts

5.
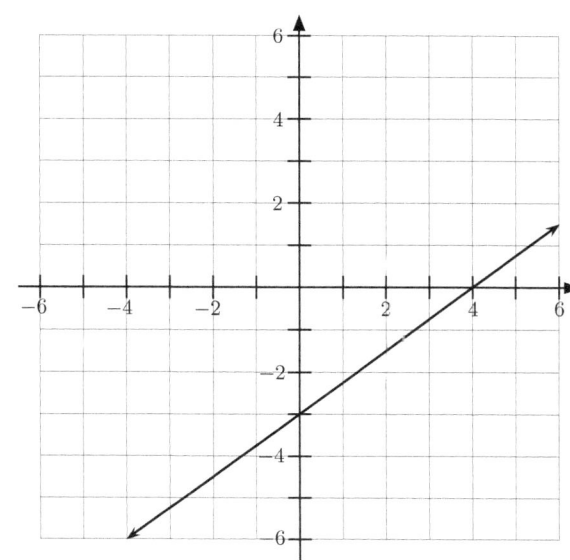

6.
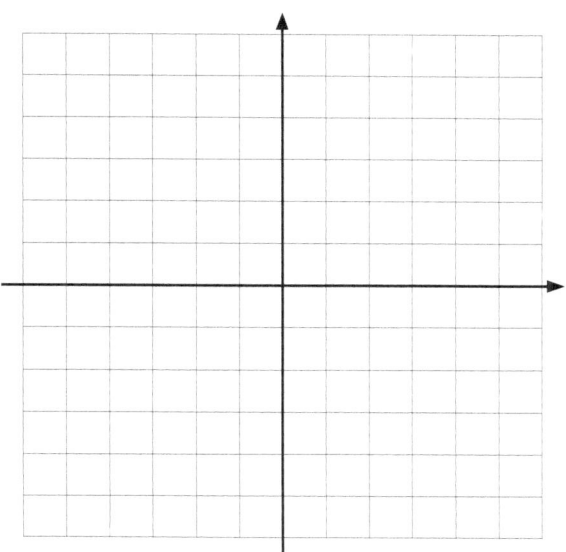

For each equation, find the x- and y-intercepts, if they exist. Then sketch the line, finding more points as necessary.

7. $y = -\frac{1}{4}x - 3$

8. $y = x - 1$

9. $y = x$

10. $y + 4x = 2$

11. $y = -1$

12. $x - y = -3$

13. $x = 1$

14. $3x + y = 0$

15. $4x + y = 5$

16. $3x + 5y = 15$

17. $y = \frac{1}{2}x$

18. $y = -x$

19. $-3x + y = 6$

20. $x - 2y = 8$

Mid-Chapter 2 Check-Up

1. Is the following ordered pair, $\left(-\frac{1}{3}, \frac{20}{7}\right)$, a solution to the equation $7y - 3x = 21$?

2. For each equation, produce a table of at least 3 solutions. Then use those points to graph the line.

   a) $3x - 2y = 6$

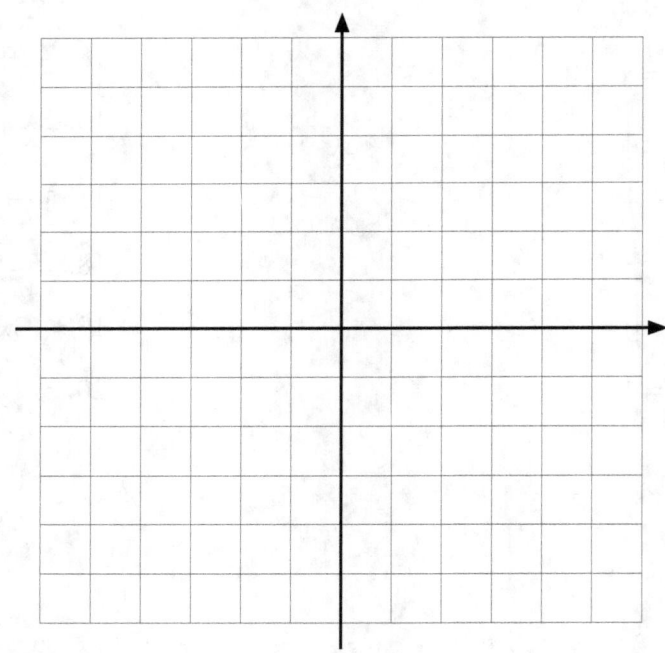

x	y

2.3 Intercepts

b) $y = -3$

x	y

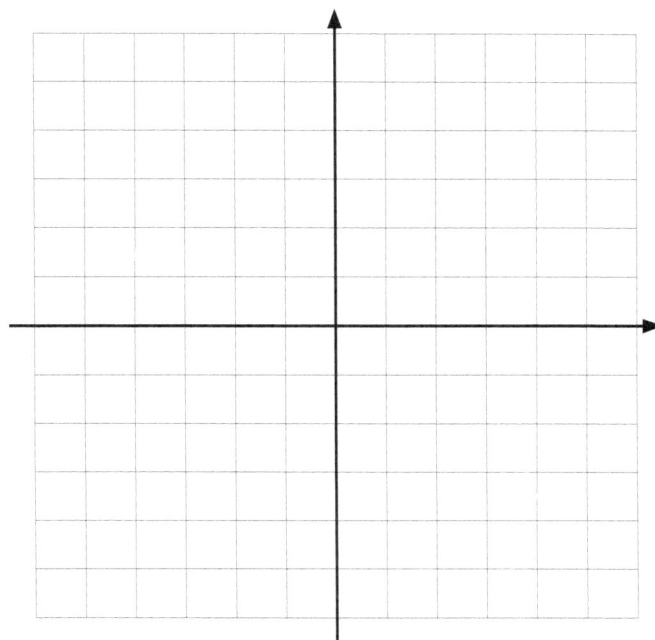

3 For each equation, find the x-intercept and y-intercept. Write your answers as ordered pairs. Then graph the line using the intercepts found.

a) $y = -x + 5$

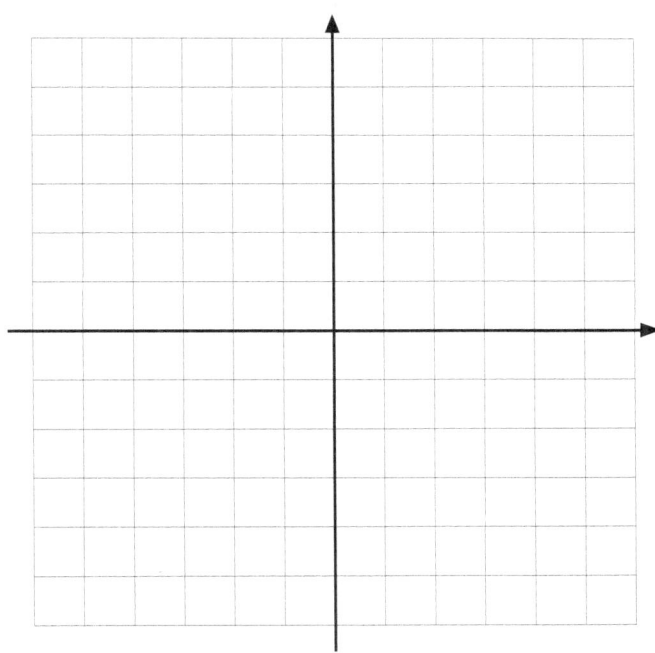

b) $2x + y = -4$

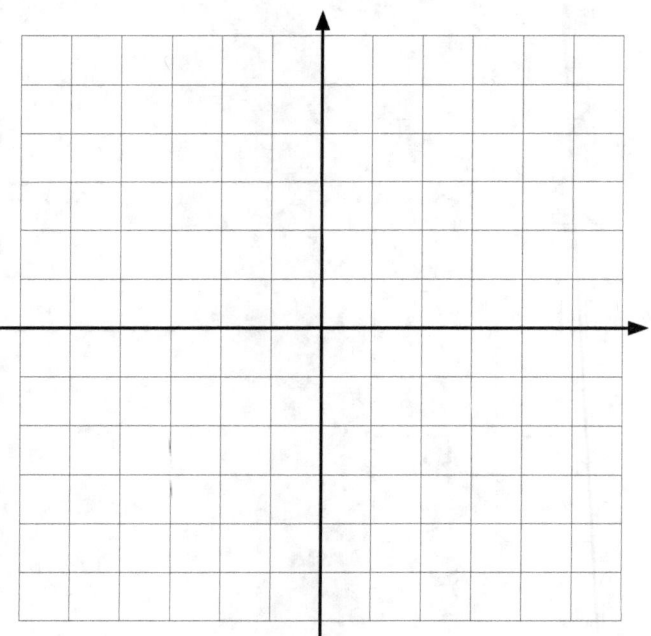

2.4 Slope

Objective: To understand the meaning of the slope of a line, find the slope given two points and use the slope to solve application problems.

What do these descriptions have in common?

- the pitch or steepness of a roof
- the grade of a road on a mountain
- the incline of a wheelchair ramp

These all refer to the steepness of a surface. The **slope** is a way to measure the steepness of a line. A line with a large positive slope like 25, is very steep. A line with a small positive slope, such as $\frac{1}{20}$, is close to flat. we also use slope to describe the direction of a line.

- A line that goes up from left to right has a **positive slope.**
- A line that goes down from left to right has a **negative slope.**

Exercise 1 **Class Example**
Determine whether the line has a positive or a negative slope.
a)   b)

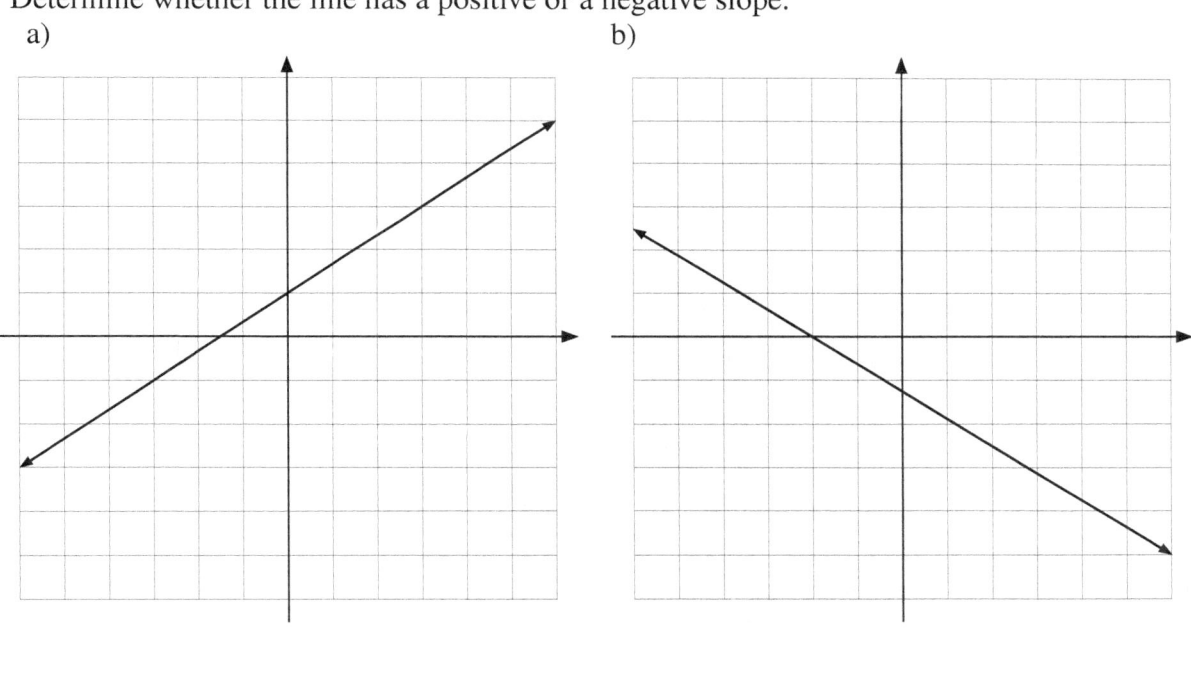

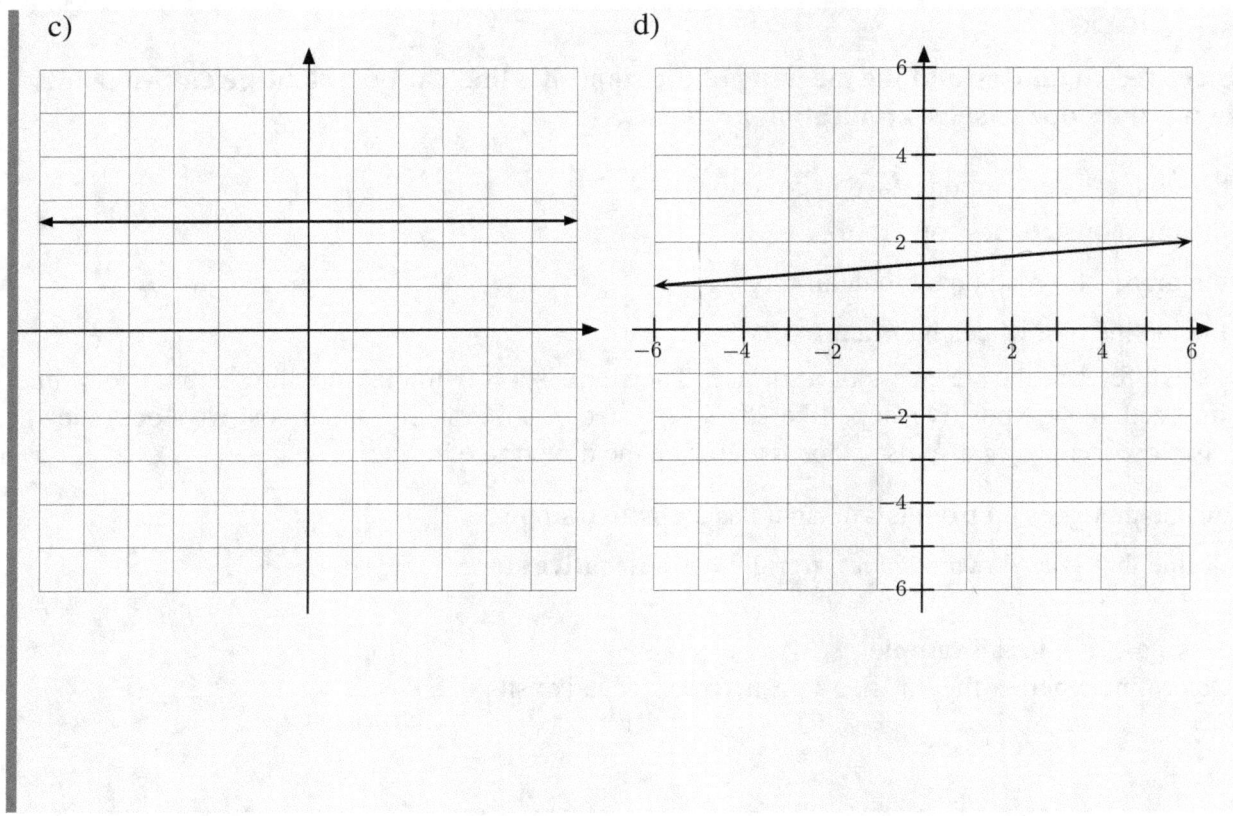

Reading the Slope from a Graph

As we measure steepness, we are interested in how fast the line rises compared to how far the line runs.

- the **rise** is the vertical change or the change in the y-coordinates
- the **run** is the horizontal change or the change in the x-coordinates

Therefore,

$$\text{slope} = \frac{\text{rise}}{\text{run}} = \frac{\text{change in y}}{\text{change in x}}$$

To compute the slope from the graph of a line, do the following.

- Pick any two points on the line. It is easier to compute the slope if the coordinates of the points are all integers.

- Count the number of units from one point to the other point in the direction of the coordinate axes.

  ◇ For the rise, the vertical change or change in y is positive when you go up and negative when you go down.

  ◇ For the run, the horizontal change or change in x is positive when you go to the right and negative when you go to the left.

2.4 Slope 129

Example 1
Find the slope of the given line.

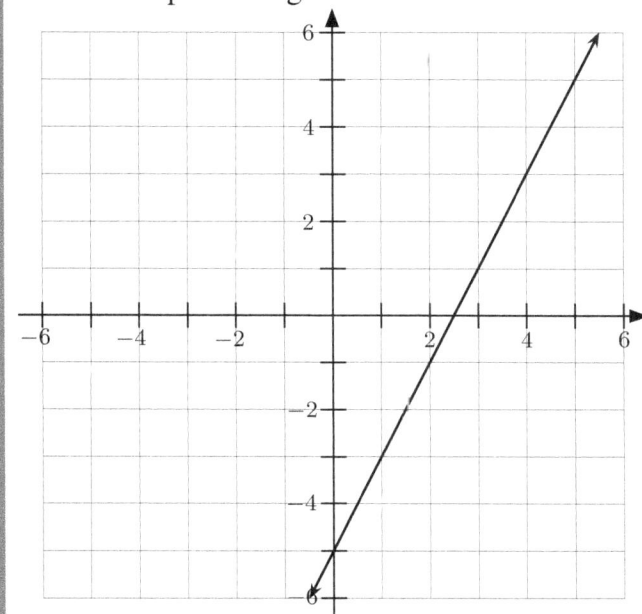

Solution.

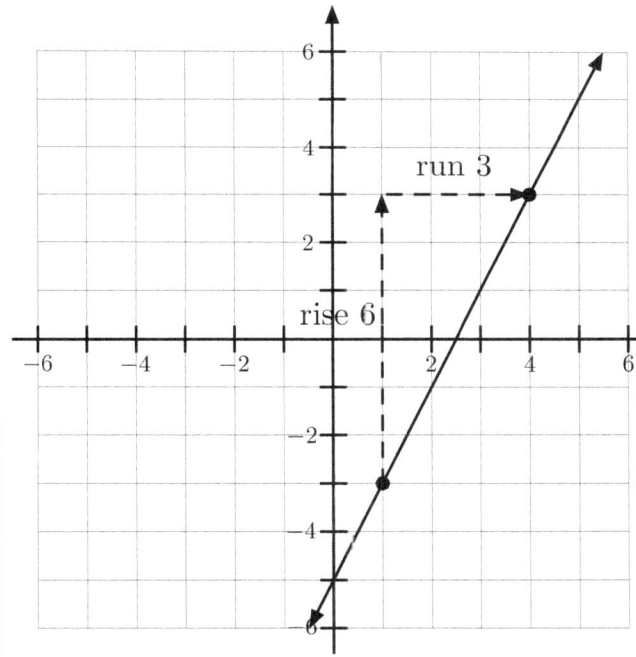

Note that the line goes up from left to right. We should have a positive slope.

Pick two points on the line. In this case, we will pick the points $(1,-3)$ and $(4,3)$. From the point $(1,-3)$, we need to go 6 units up and 3 units right in order to get to the point $(1,-3)$. This is a rise of 6 and a run of 3.

The slope, written as a fraction, is $\frac{\text{rise}}{\text{run}} = \frac{6}{3} = 2$.

Therefore, the slope of the line is 2.

Example 2
Find the slope of the given line.

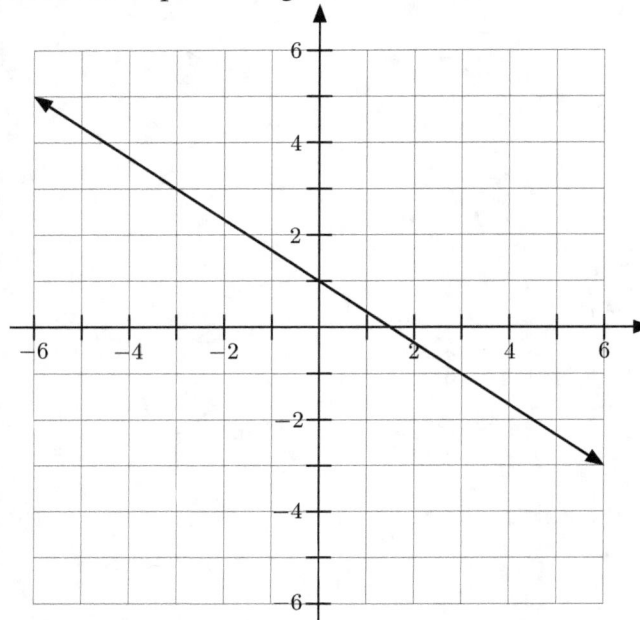

Solution.

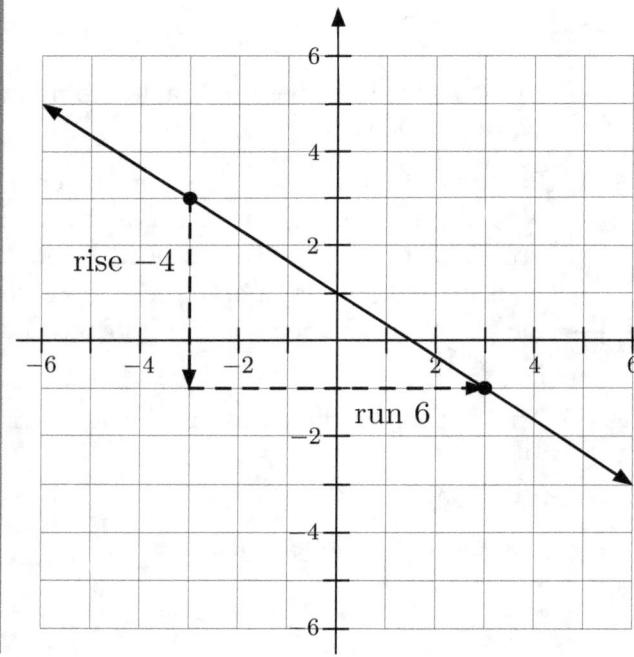

Note that the line goes down from left to right. We should have a negative slope.

Pick two points on the line. In this case, we will pick the points $(-3, 3)$ and $(3, -1)$. From the point $(-3, 3)$, we need to go 4 units down and 6 units right in order to get to the point $(3, -1)$.

This is a rise of -4 and a run of 6.

The slope, written as a fraction, is $\frac{\text{rise}}{\text{run}} = \frac{-4}{6} = \frac{-2}{3}$.

Therefore, the slope of the line is $-\frac{2}{3}$.

2.4 Slope

Exercise 2 Class Example
Choose two points on the line and use them to find the slope of the line.

a)
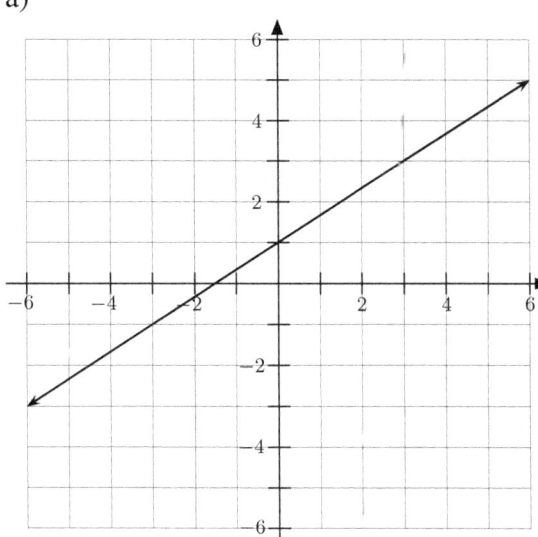

b)
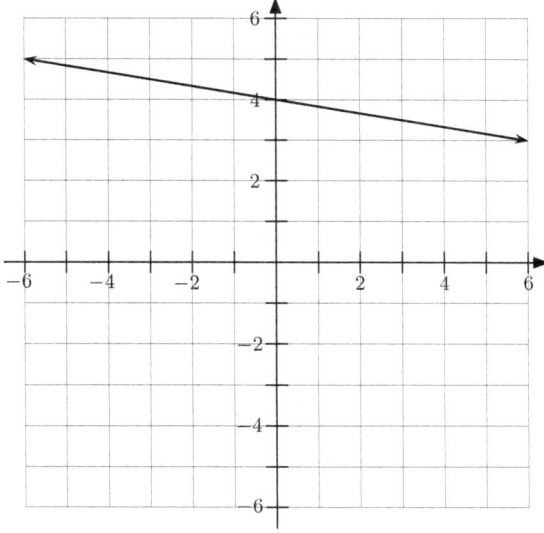

Exercise 3 You Try
Choose two points on the line and use them to find the slope of the line.

a)
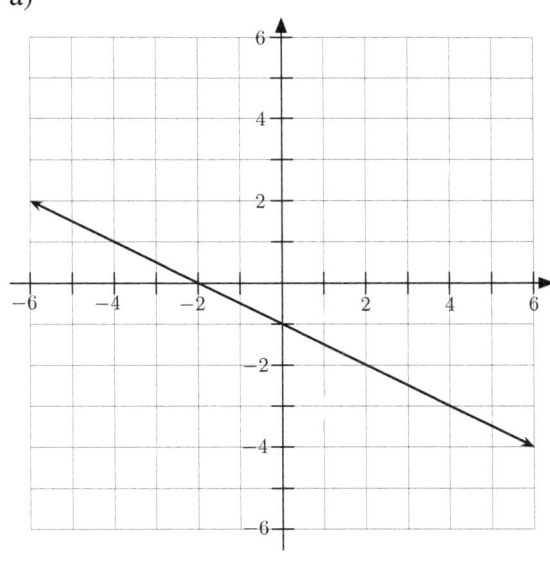

b)
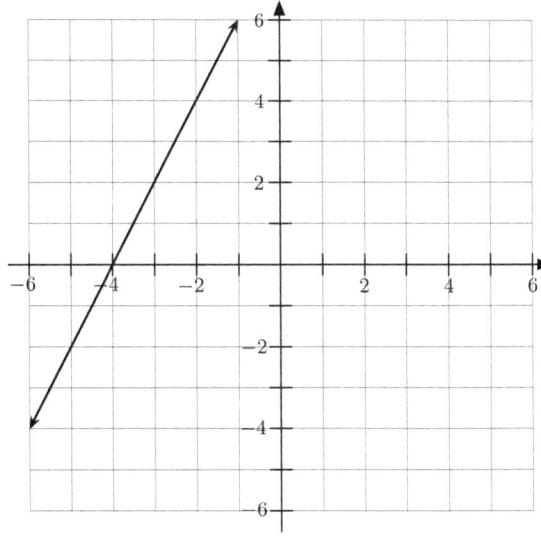

Which two points did you choose? Notice that **any** two points on a line can determine the slope of the line. The slope will be the same no matter which points from the line are chosen.

Slope of Horizontal and Vertical Lines

We need to be aware of two special kinds of lines that have unique slopes. Recall that a horizontal line contains points with the same y-coordinate. What does that mean regarding its slope? Given two points on a horizontal line, the rise will always be 0 for any run. In other words, slope = $\frac{0}{\text{run}}$. Therefore, a horizontal line has a slope of 0.

Similarly, a vertical line contains point with the same x-coordinate. Given two points on a vertical line, for any rise, the run will always be 0. In other words, slope = $\frac{\text{rise}}{0}$. However, dividing by 0 is not possible. Therefore, a vertical line has an undefined slope.

Example 3 Find the slope of the following lines.
a)   b)

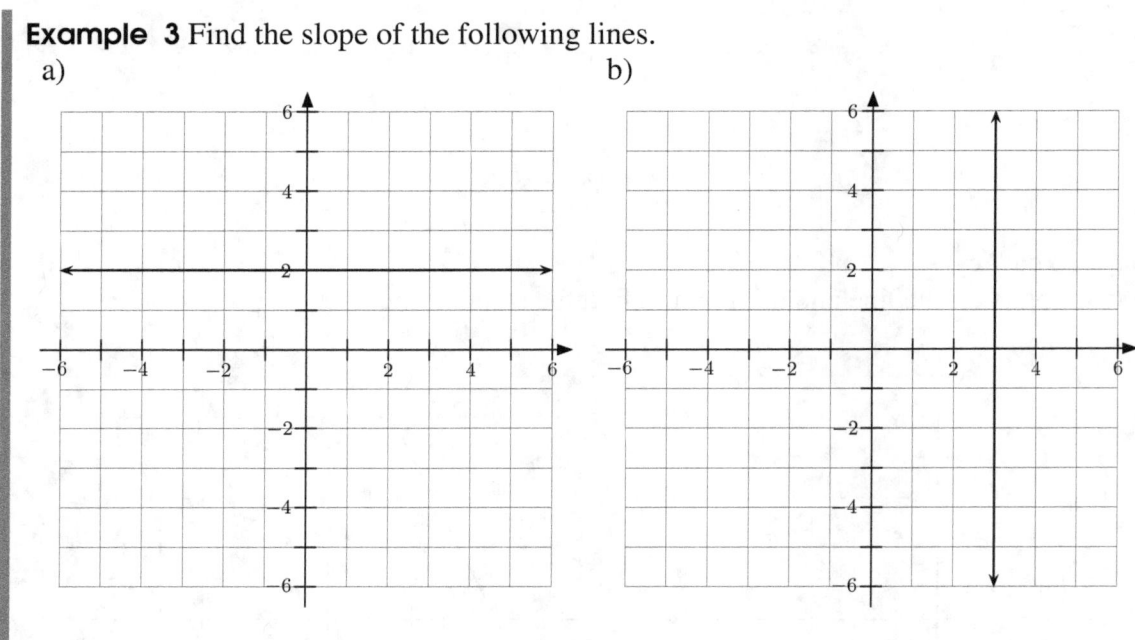

2.4 Slope

Solution.

a)

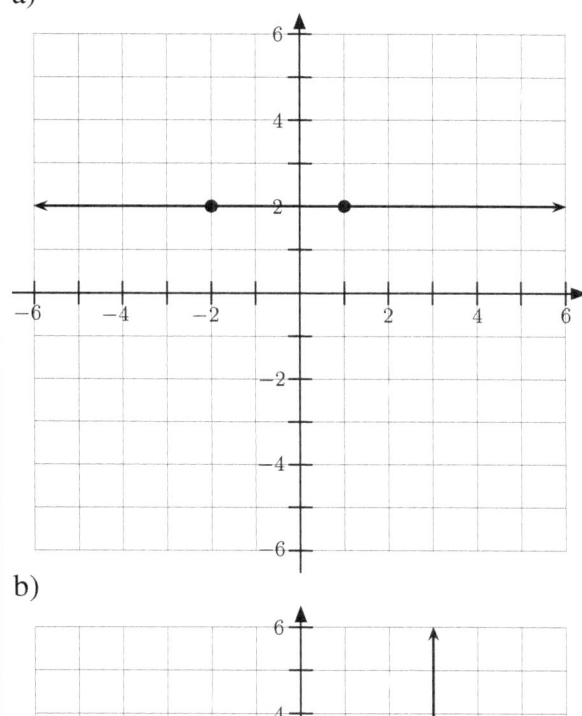

Pick two points on the line. In this case, we will pick the points $(-2, 2)$ and $(1, 2)$.

From the point $(-2, 2)$, we don't go up but we only go to the right 3 units to get to the point $(1, 2)$. That means, the rise is 0 and the run is 3 which gives us

slope = $\frac{0}{3} = 0$.

This line, and all horizontal lines, have a slope of 0.

b)

Pick two points on the line. In this case, we will pick the points $(3, 3)$ and $(3, -2)$.

From the point $(3, 3)$, we go down 5 units and we are now at the point $(3, -2)$. That means, the rise is -5 and the run is 0.

The slope = $\frac{-5}{0}$ which is undefined.

This line, and all vertical lines, have an undefined slope.

As you can see, there is a big difference between having a zero slope and having an undefined slope. Remember, slope is a measure of steepness. A horizontal line is not steep at all. In fact, it is flat. Therefore, a horizontal line has zero slope.

On the other hand, a vertical line cannot get any steeper. It is so steep that there is no number large enough to express how steep it is. Therefore, a vertical line has an undefined slope.

Slopes of Vertical and Horizontal Lines

- A horizontal line has a slope of 0.
- A vertical line has an undefined slope.

Exercise 4 Class Example
Find the slope of each of the following lines.
 a) b)

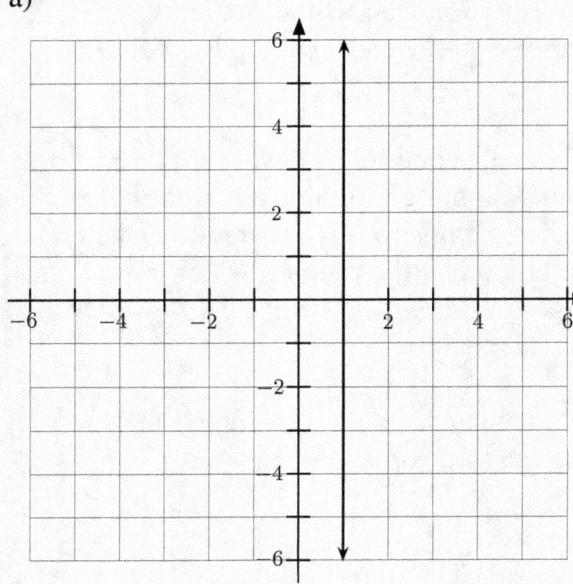 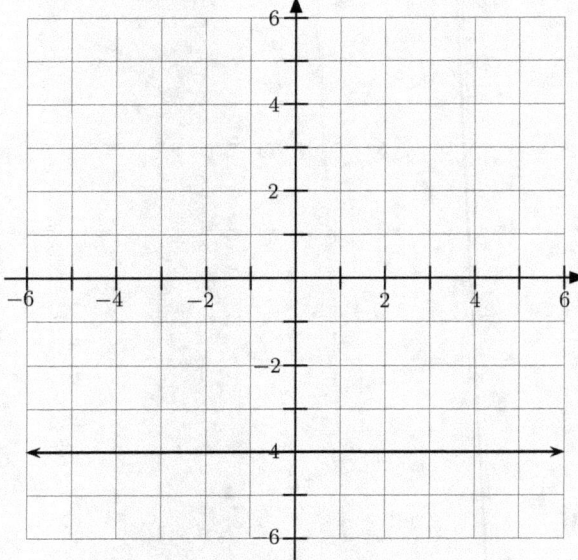

Exercise 5 You Try
Find the slope of each of the following lines.
 a) b)

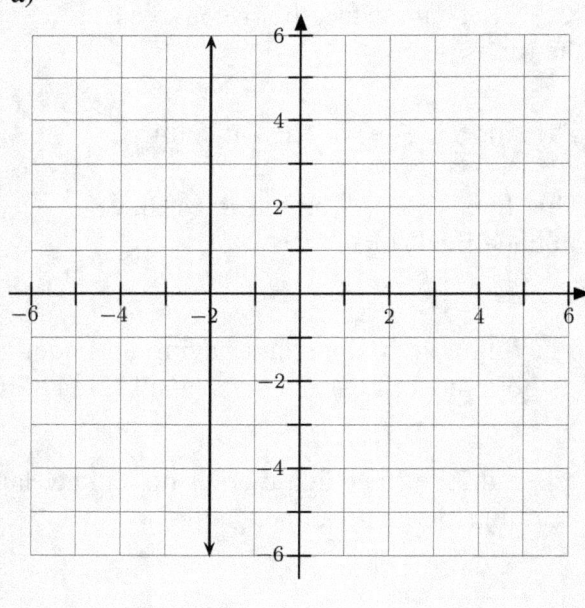 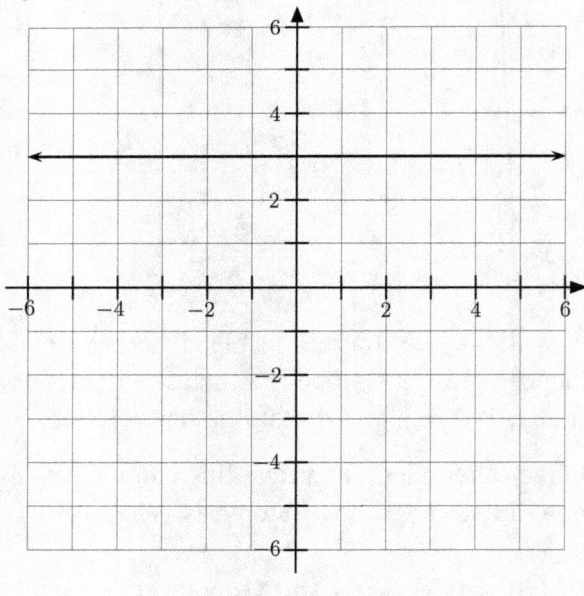

Slope Formula

In addition to reading a graph, we can find the slope of a line through any two points using the **slope formula**.

> **Slope Formula**
>
> Given any two points (x_1, y_1) and (x_2, y_2), the slope of the line containing these points is given by the following formula.
>
> $$\text{Slope} = m = \frac{y_2 - y_1}{x_2 - x_1}$$

Example 4 Use the slope formula to find the slope of the line containing the points $(-4, 3)$ and $(2, -9)$.

Solution.
Let $(x_1, y_1) = (-4, 3)$ and $(x_2, y_2) = (2, -9)$. Substitute the information into the slope formula to find the slope of the line containing these points.

$$m = \frac{y_2 - y_1}{x_2 - x_1} \quad \text{Substitute values}$$

$$= \frac{-9 - 3}{2 - (-4)} \quad \text{Perform indicated operation}$$

$$= \frac{-12}{6} \quad \text{Simplify fraction}$$

$$= -2 \quad \text{Slope of the line}$$

What if we let $(x_1, y_1) = (2, -9)$ and $(x_2, y_2) = (-4, 3)$. Let us compute the slope of the line.

$$m = \frac{y_2 - y_1}{x_2 - x_1} \quad \text{Substitute values}$$

$$= \frac{3 - (-9)}{-4 - 2} \quad \text{Perform indicated operation}$$

$$= \frac{12}{-6} \quad \text{Simplify fraction}$$

$$= -2 \quad \text{Slope of the line}$$

We see that it does not matter which point we call (x_1, y_1) and which point we call (x_2, y_2). The slope of the line containing the points is still the same.

Exercise 6 Class Example
Use the slope formula to find the slope of the line containing the given points.

a) $(4, 6)$ and $(2, -1)$

b) $(-4, -1)$ and $(-4, -5)$

Exercise 7 Class Example
Use the slope formula to find the slope of the line containing the given points.

a) $(3, 1)$ and $(-2, 6)$

b) $(6, -5)$ and $(-2, 7)$

Using the slope formula, we can find the missing coordinate if we know what the slope is. Let us take a look at an example.

Example 5 A line with a slope of -3 passes through the points $(5, -1)$ and $(2, y)$. Find the value of y.

Solution.
Let $(x_1, y_1) = (5, -1)$ and $(x_2, y_2) = (2, y)$. Substitute the information into the slope formula to

2.4 Slope

solve for y.

$$m = \frac{y_2 - y_1}{x_2 - x_1} \qquad \text{Substitute values}$$

$$-3 = \frac{y - (-1)}{2 - 5} \qquad \text{Perform indicated operation}$$

$$-3 = \frac{y + 1}{-3} \qquad \text{Multiply each side by} -3$$

$$-3(-3) = \frac{y + 1}{-3}(-3) \qquad \text{Perform indicated operation}$$

$$9 = y + 1 \qquad \text{Subtract 1 from each side}$$

$$8 = y \qquad \text{Our Solution}$$

Verify that the solution yields a slope of -3 by letting $(x_1, y_1) = (5, -1)$ and $(x_2, y_2) = (2, 8)$.

$$-3 \stackrel{?}{=} \frac{8 - (-1)}{2 - 5}$$

$$-3 = \frac{9}{-3} \quad \checkmark$$

Exercise 8 Class Example

A line with a slope of $\frac{2}{5}$ passes through the points $(-3, 2)$ and $(x, 6)$. Find the value of x.

Exercise 9 You Try
A line with a slope of 2 passes through the points $(1, y)$ and $(5, 4)$. Find the value of x.

Using a Point and the Slope to Find Another Point

Given a point p, we can also use the slope formula to find other points that make a desired slope from p. Lets look at an example.

If we are given a slope and a point, we have enough information to find other points on the same line. Lets take a look at an example.

Example 6 Given the point $(2, 1)$, find two other points that make a slope of $\frac{4}{3}$.

Solution.

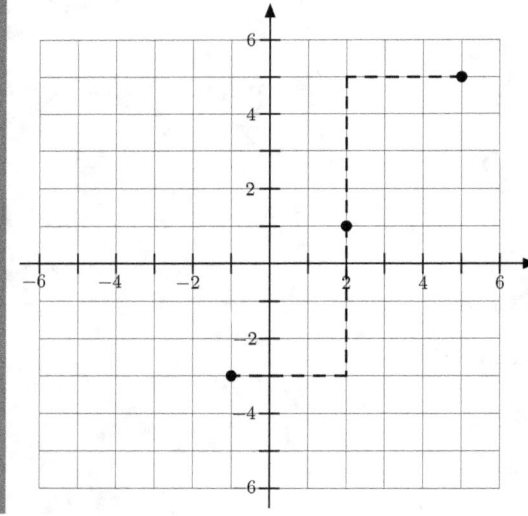

A slope $= m = \frac{4}{3}$, means that for every rise of 4, we need a corresponding run of 3. Starting at the point $(2, 1)$, if we go up 4 units and right 3 units, we arrive at the point $(5, 5)$.

We can use equivalent fractions to rewrite the slope, $m = \frac{4}{3} = \frac{-4}{-3}$. That means that for every rise of -4, we need a corresponding run of -3. Starting at $(2, 1)$, if we go down 4 units and left 3 units, we arrive at the point $(-1, -3)$.

Exercise 10 Class Example
Given the point $(1,0)$, find two other points that make a slope of -3.

Exercise 11 You Try

a) Given the point $(-3, 2)$, find two other points that make a slope of $-\dfrac{5}{2}$.

b) Given the point $(2, 1)$, find two other points that make a slope of 0.

Applications of the Slope

Another way to describe slope is as the **rate of change**. In the context of an application, it is important to include the units when providing the answer.

Example 7 On a drive from Seattle to San Francisco, you notice that after 2 hours, you had driven 124 miles. After 5 hours, you had driven 310 miles.

a) What is the rate of change of distance with respect to time?

b) Interpret the meaning of the rate of change in the context of the situation.

Solution.

a) Let us start by writing the given data as 2 points. Since we want rate of change of distance with respect to time, we select time as the first coordinate and distance as the second coordinate. The points will look like this: (time, distance). Therefore, our two points are $(2, 124)$ and $(5, 310)$. Substitute these values into the slope formula.

$$m = \frac{y_2 - y_1}{x_2 - x_1} \quad \text{Substitute values into slope formula}$$

$$= \frac{310 - 124}{5 - 2} \quad \text{Perform indicated operation}$$

$$= \frac{186}{3} \quad \text{Simplify fraction}$$

$$= 62 \quad \text{Slope}$$

What are the units? The y-coordinates are in miles and the x-coordinates are in hours. So the rate of change is 62 miles/hour.

b) Within the 3 hour period, you have been driving at a rate of 62 miles/hour.

Exercise 12 Class Example

The clarity of Lake Tahoe is measured several times a year by scientists at UC Davis. In 2016, the lowest clarity level was reported to be 44.3 feet. In 1997, the lowest clarity of the lake was reported to be 64.1 feet.
(Source: http://news.ucdavis.edu/search/news_details.lasso?id=11171)

a) What is the rate of change of the depth of clarity with respect to time? Round your answer to the nearest tenth and be sure to include units in your final answer.

b) Interpret the meaning of the rate of change in the context of the situation.

2.4 Slope

Exercise 13 You Try

Sue's new car holds 11 gallons of gas and she just filled up her tank. After driving a distance of 256 miles, she approximates that she used up 8.5 gallons of gas.

a) What is the rate of change of the gallons of gas with respect to distance? Round your answer to the nearest tenth and be sure to include units in your final answer.

b) Interpret the meaning of the rate of change in the context of the situation.

2.4: Exercises

Find the slope of each line.

1.

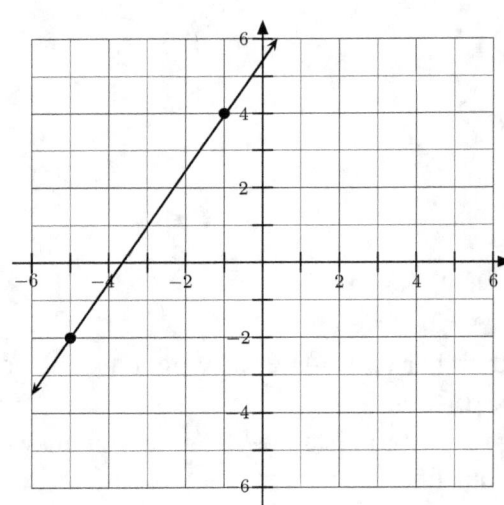

3.

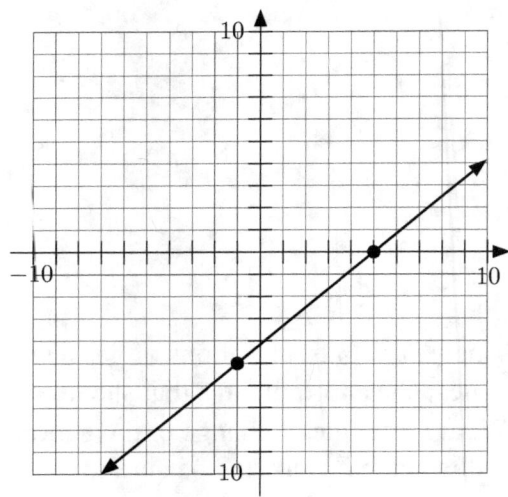

2.

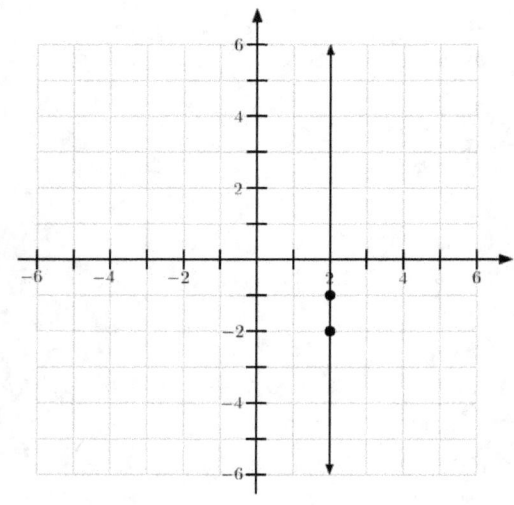

4.

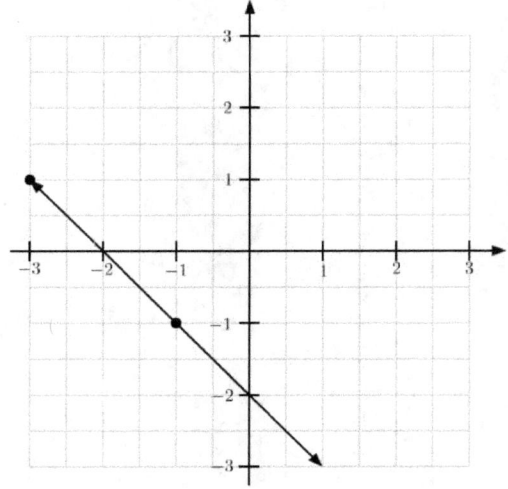

2.4 Slope 143

5.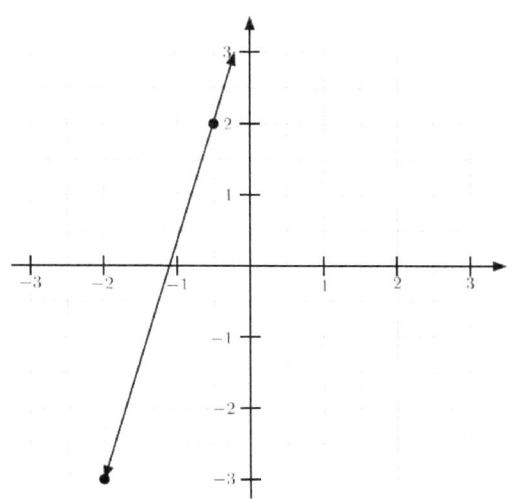

6.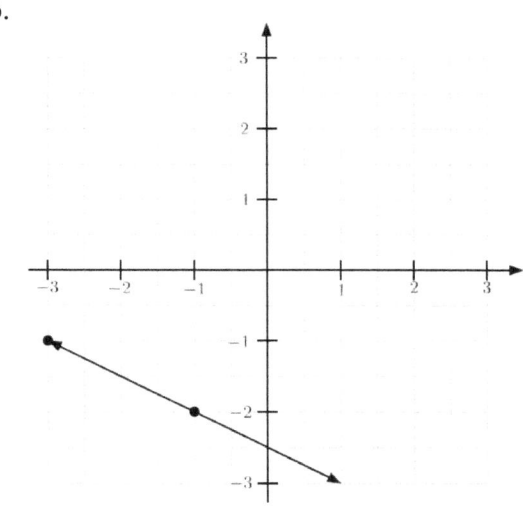

7.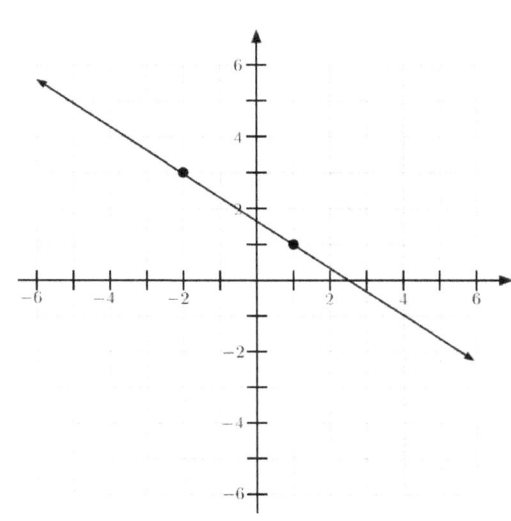

8.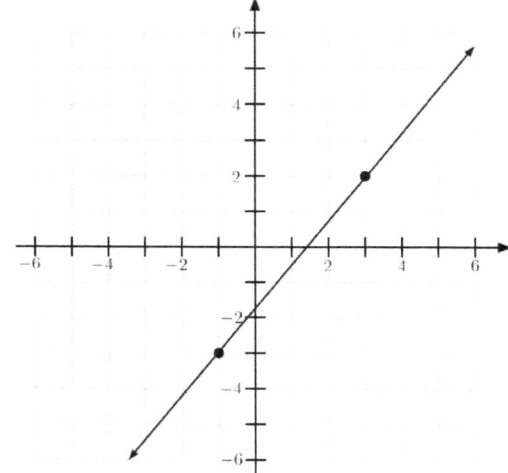

9.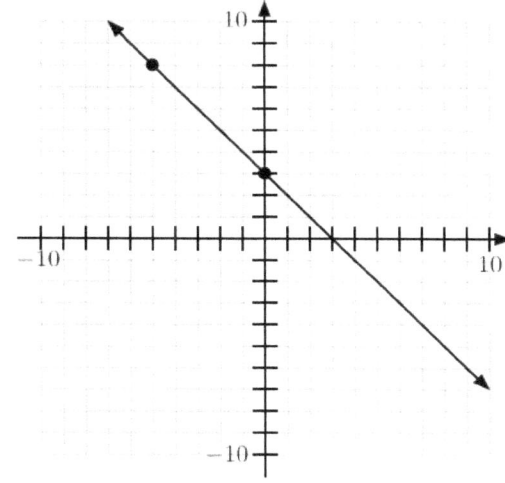

10.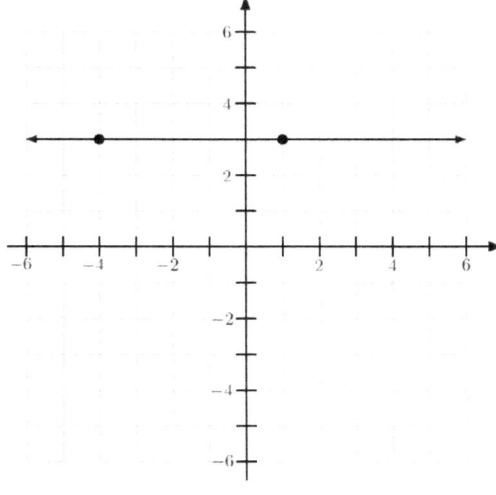

Find the slope of the line through each pair of points.

11. $(-2, 10), (-2, 15)$

12. $(1, 2), (-6, -14)$

13. $(-15, 10), (16, -7)$

14. $(13, -2), (7, 7)$

15. $(10, 18), (-11, -10)$

16. $(-3, 6), (-20, 13)$

17. $(-16, -14), (11, -14)$

18. $(13, 15), (2, 10)$

19. $(-4, 14), (-16, 8)$

20. $(9, -6), (-7, -7)$

Find the value of x or y so that the line through the points has the given slope.

21. $(-8, y)$ and $(-1, 1)$; slope: $\dfrac{6}{7}$

22. $(8, y)$ and $(-2, 4)$; slope: $-\dfrac{1}{5}$

23. $(-2, y)$ and $(2, 4)$; slope: $\dfrac{1}{4}$

24. $(2, -5)$ and $(3, y)$; slope: 6

Using the given point, find another point that makes the provided slope.

25. $(6, 14)$; $m = \dfrac{1}{2}$

26. $(5, -2)$; undefined slope

27. $(-3, 8)$; $m = -4$

28. $(0, 2)$; $m = 0$

29. $(-4, 7)$; $m = 5$

30. $(1, 0)$; $m = -\dfrac{3}{2}$

Applications

31. The recommended dosage of a certain medicatiuon for a 50 lb. dog is 25 mL. For a 16 lb. dog, the dosage recommendation is 8mL.

    (a) What is the rate of change of the dosage with respect to weight?

    (b) What would the recommended dosage be for a 35 lb. dog?

32. In January 2012, the average 2 bedroom home in Seattle was worth $308,000. In January 2015, that value was $410,000. What is the rate of change of home prices with respect to time? (Source: www.zillow.com/home-values/)

33. In 2004, Washington State recycled or diverted 6,223,974 tons of waste. In 2013, Washington State recycled or diverted 7,961,040 tons. What is the rate of change of recycled or diverted waste with respect to time? (Source: www.ecy.wa.gov/beyondwaste/bwprog_swGenRec.html)

2.4 Slope

Rescue Roody!

34. Roody needs to find the slope between these points: $(3,4)$ and $(1,-2)$. He cannot seem to find the correct slope. Help him determine what he did wrong.

$$m = \frac{y_2 - y_1}{x_2 - x_1}$$
$$= \frac{-2 - 4}{3 - 1}$$
$$= \frac{-6}{2}$$
$$= -3$$

Challenge

For each table, determine if the data represents a direct linear variation or not.

35.

x	y
2	3
5	8
11	18
14	23

36.

x	y
1	1
2	4
3	9
4	16

2.5 Slope-Intercept Form

Objective: To understand the slope-intercept form and use it to graph a line

If we know the slope of a line, m, and we know a point on that line, (x_1, y_1), then we can use the slope formula to determine all other points that belong to the line. Remember that any point (x, y) will belong to this line if it makes a slope of m with the given point (x_1, y_1). In this section, we will work with the case where the point is the y-intercept.

Example 1 A line has slope, $m = \dfrac{2}{3}$ and y-intercept, $(0, 1)$. Find the equation of the line.

Solution.
A point (x, y) will be on the line if it makes a slope of $m = \dfrac{2}{3}$ with the y-intercept $(0, 1)$. Use the slope formula to obtain the equation of the line.

$$\dfrac{y_2 - y_1}{x_2 - x_1} = m \qquad \text{Slope formula}$$

$$\dfrac{y - 1}{x - 0} = \dfrac{2}{3} \qquad \text{Substitute values}$$

$$\dfrac{y - 1}{x} = \dfrac{2}{3} \qquad \text{Solve for y}$$

$$(x)\dfrac{y - 1}{x} = \dfrac{2}{3}(x) \qquad \text{Multiply each side by x}$$

$$y - 1 = \dfrac{2}{3}x \qquad \text{Add 1 to each side}$$

$$y = \dfrac{2}{3}x + 1 \qquad \text{Our equation of the line}$$

From the equation of the line, $y = \dfrac{2}{3}x + 1$, we notice the following.

- the coefficient of x is the slope of the line, ie, $m = \dfrac{2}{3}$
- the equation's constant value is the y-coordinate of the y-intercept. Let us see why. Since the y-intercept always has a 0 for the x-coordinate, lets set $x = 0$.

$$y = \dfrac{2}{3}(0) + 1$$

$$y = 0 + 1$$

$$y = 1$$

This is the constant value of the linear equation. It is also the y-coordinate of the y-intercept.

2.5 Slope-Intercept Form

Slope-Intercept Form
The equation, $y = mx + b$, is called the **Slope-Intercept Form** of the line where

- m is the slope
- $(0, b)$ is the y-intercept

Example 2 Identify the slope and y-intercept of the line $y = 2x - 3$.

Solution.
This equation is in slope-intercept form, where $m = 2$ and $b = -3$. Therefore, the slope of the line is $m = 2$ and the y-intercept is the point $(0, -3)$.

Exercise 1 **Class Example**
Identify the slope and y-intercept of the line $y = -\frac{3}{5}x + 7$.

Exercise 2 **You Try**
Identify the slope and y-intercept of the line $y = x - 8$.

Example 3 Identify the slope and y-intercept of the line $4x - y = 9$.

Solution.
We must first transform the equation of the line into slope-intercept form by solving for y.

$$4x - y = 9$$
$$\underline{-4x \qquad -4x} \qquad \text{Subtract 4x from each side}$$
$$-y = -4x + 9$$
$$\frac{-y}{-1} = \frac{-4x}{-1} + \frac{9}{-1} \qquad \text{Divide each term by } -1$$
$$y = 4x - 9 \qquad \text{Equation of line in slope-intercept form}$$

Now that we have the equation of the line in slope-intercept form, we can see that the slope is $m = 4$ and the y-intercept is the point $(0, -9)$.

Exercise 3 Class Example
For each of the following lines, identify the slope and y-intercept.

a) $x + y = 2$ 　　　　　　　　　　　　b) $6y = 3x$

Exercise 4 Ypu Try
For each of the following lines, identify the slope and y-intercept.

a) $3x + y = 1$ 　　　　　　　　　　　　b) $5x - y = 7$

2.5 Slope-Intercept Form

Graphing Lines Using the Slope and the y-Intercept

When graphing lines using the slope and y-intercept, solve for the variable y, so the equation of the line is in slope-intercept form, $y = mx + b$. From the equation, identify the slope and y-intercept. To find another point on the line, start with the y-intercept. Use the slope to rise (up or down) and run (left or right) to get a second point on the line.

Be careful with the sign of the slope. Since the slope can be written as a fraction, which is a quotient of 2 integers, a positive fraction means that 2 positive integers or 2 negative integers are being divided. While a negative fraction means that one positive and one negative integer are being divided. Let us take a look at how that translates in graphing a line.

- If the slope is positive, say $m = \frac{2}{3}$, we can move in the positive y-direction and positive x-direction from the y-intercept to get the second point. Since $m = \frac{2}{3} = \frac{-2}{-3}$, we can also move in the negative y-direction and negative x-direction to get the second point. Depending on which direction we move, our second point will be different. However, it will still be a point on the line.

- If the slope is negative, say $m = -\frac{2}{3} = \frac{-2}{3}$, we can move in the negative y-direction and positive x-direction from the y-intercept to get the second point. Since $m = -\frac{2}{3} = \frac{2}{-3}$, we can also move in the positive y-direction and negative x-direction to get the second point. Depending on which direction we move, our second point will be different. However, it will still be a point on the line.

Example 4 Identify the slope and the y-intercept of the line $y = -\frac{4}{3}x + 3$ and use the information to graph the line.

Solution.
Since the equation of the line is in slope-intercept form, we can easily determine the slope, $m = -\frac{4}{3} = \frac{-4}{3} = \frac{4}{-3}$ and y-intercept, $(0, 3)$.

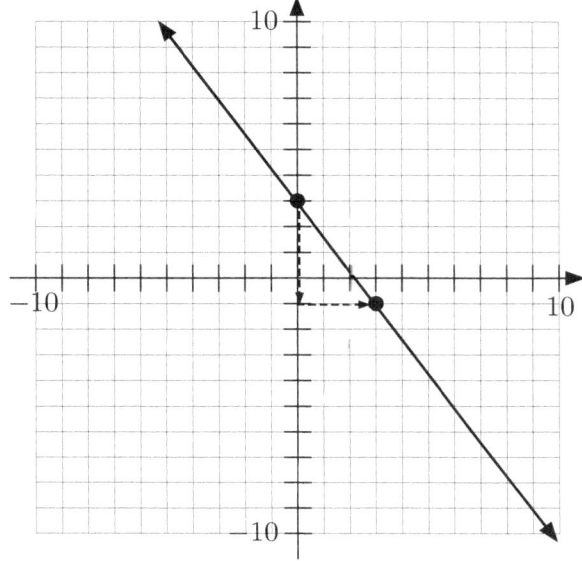

To graph the line using the slope, $m = \frac{-4}{3}$, we start at the y-intercept, $(0, 3)$.

From there, we go down 4 units and then go to the right 3 units.

We will arrive at the point $(3, -1)$.

To draw the line we connect the points $(0, 3)$ and $(3, -1)$ and continue the line in each direction.

Chapter 2. Graphing Linear Equations

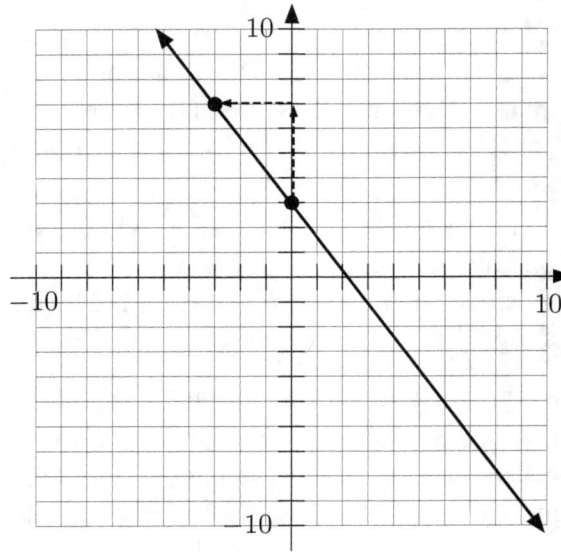

In the last example we could have also used the slope m as $m = \frac{4}{-3}$.
To graph the line, start at the y-intercept, $(0, 3)$.
From there, go up 4 units and then go left 3 units. We will arrive at the point $(-3, 7)$.
Connect the points $(0, 3)$ and $(-3, 7)$ and continue the line in each direction.
You will see that for the slope $m = -\frac{4}{3}$, it does not matter whether we go down first and then go right or go up first and then go left. The line goes through all the points we found.

Example 5 Identify the slope and the y-intercept of the line $3x - 2y = 10$ and use the information to graph the line.
Solution.
We need to rewrite the equation in slope-intercept form by solving for y.

$$3x - 2y = 10$$
$$\underline{-3x = -3x} \qquad \text{Subtract 3x from each side}$$
$$-2y = -3x + 10$$

$$\frac{-2y}{-2} = \frac{-3x}{-2} + \frac{10}{-2} \qquad \text{Divide each term by } -2$$

$$y = \frac{3}{2}x - 5 \qquad \text{Equation of the line in slope-intercept form}$$

We can now determine the slope, $m = \frac{3}{2}$ and y-intercept, $(0, -5)$.

To graph the line using the slope, $m = \frac{3}{2}$, start at the y-intercept, $(0, -5)$. From there, go up 3 units and then go right 2 units. We will arrive at the point $(2, -2)$. Connect the points $(0, -5)$ and $(2, -2)$ and continue the line in each direction.

2.5 Slope-Intercept Form

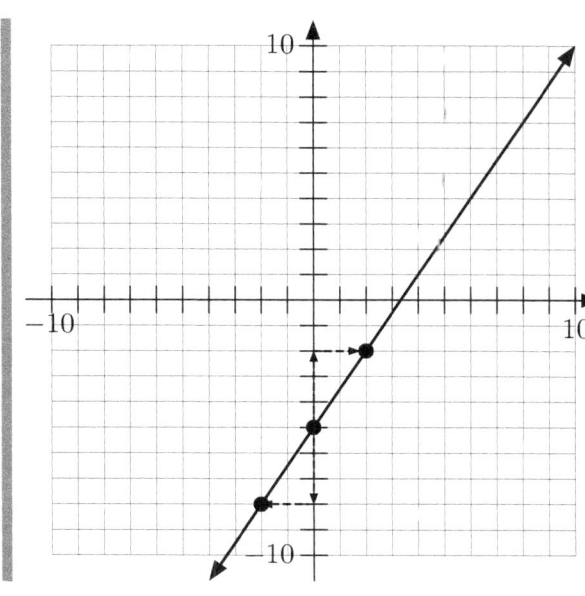

We can also use the slope, $m = \frac{-3}{-2}$.
To graph the line, start at the y-intercept $(0, -5)$.
From there, go down 3 units and then go left 2 units. We will arrive at the point $(-2, -8)$.
Connect the points $(0, -5)$ and $(-2, -8)$ and continue the line in each direction.

As before, it does not matter whether we go up first and then go right or go down first and then go left. The line goes through all the points we found.

Exercise 5 **Class Example**
Identify the slope and the y-intercept for each of the given lines and use the information to graph each line.

a) $x - y = 3$

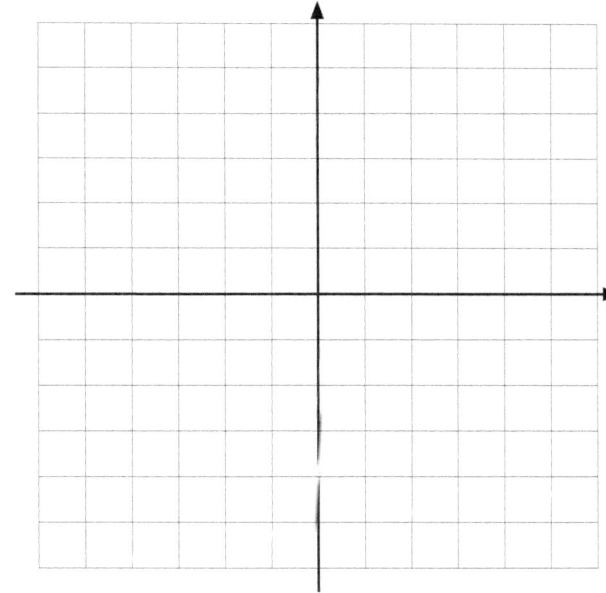

b) $3x + 4y = 8$

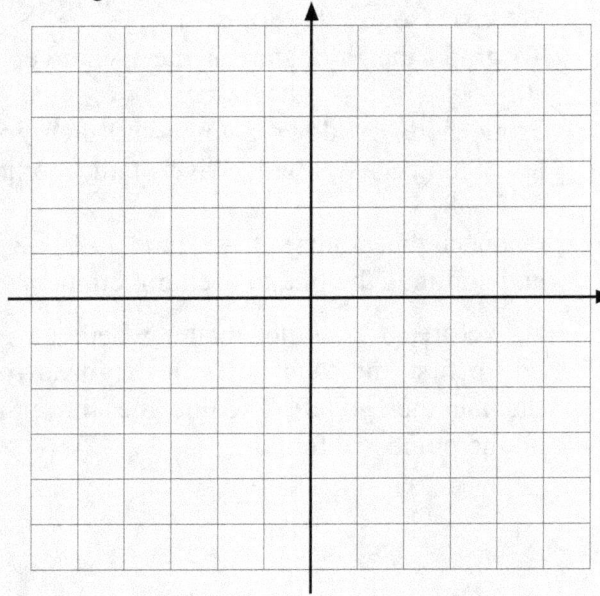

Exercise 6 **You Try**
Identify the slope and the y-intercept for each of the given lines and use the information to graph each line.

a) $x + 2y = 0$

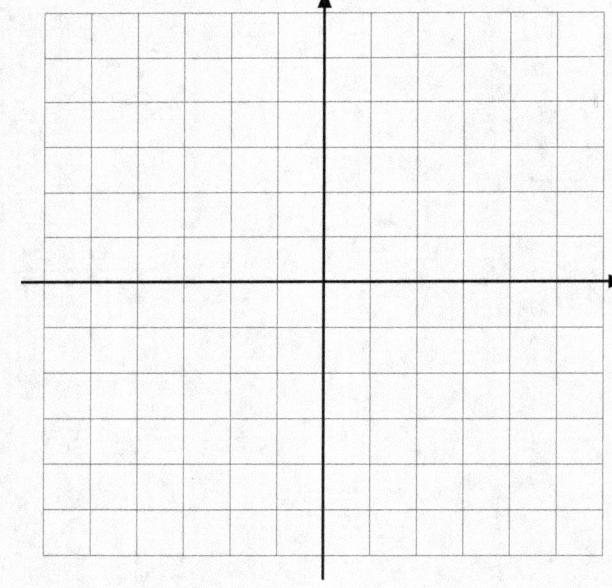

2.5 Slope-Intercept Form

b) $2x - 3y = 9$

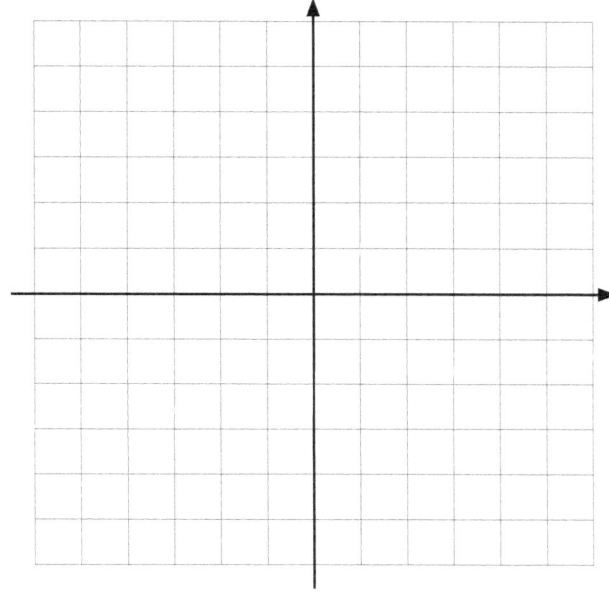

Finding the Equation of a Line from a Graph

To find the equation of a line in $y = mx + b$ form, given the graph of a line, first identify the y-intercept. The y-coordinate of that point will represent b. Then find two points on the line and use them to find the slope, m. This can be done by using the slope formula or counting how many units to rise and how many units to run. Substitute m and b into the equation, $y = mx + b$ to get the equation of the line in slope-intercept form.

Example 6 Find the equation of the line using the given graph. Write your answer in slope-intercept form.

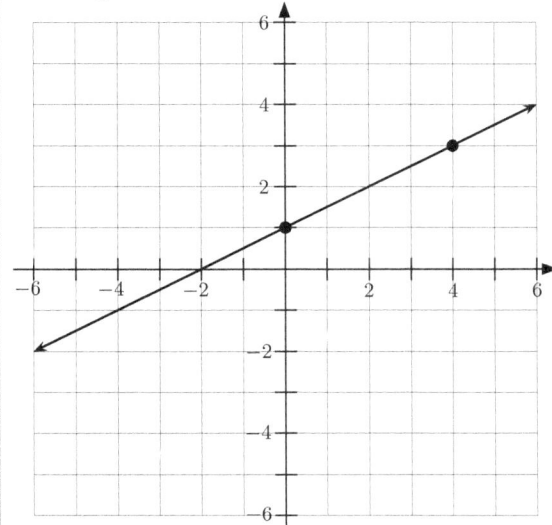

Solution.

We can read from the graph that the y-intercept is at the point, $(0,1)$. That means, $b = 1$. A second point on the graph is at $(4,3)$. Using those two points and the slope formula, we can calculate the slope, m.

$$m = \frac{y_2 - y_1}{x_2 - x_1}$$ Substitute values

$$= \frac{3-1}{4-0}$$ Perform the indicated operation

$$= \frac{2}{4}$$ Simplify fraction

$$= \frac{1}{2}$$ Slope of the line

The equation of the line in slope-intercept form is $y = \frac{1}{2}x + 1$.

Exercise 7 Class Example
Find the equation of the line using the given graph. Write your answer in slope-intercept form.

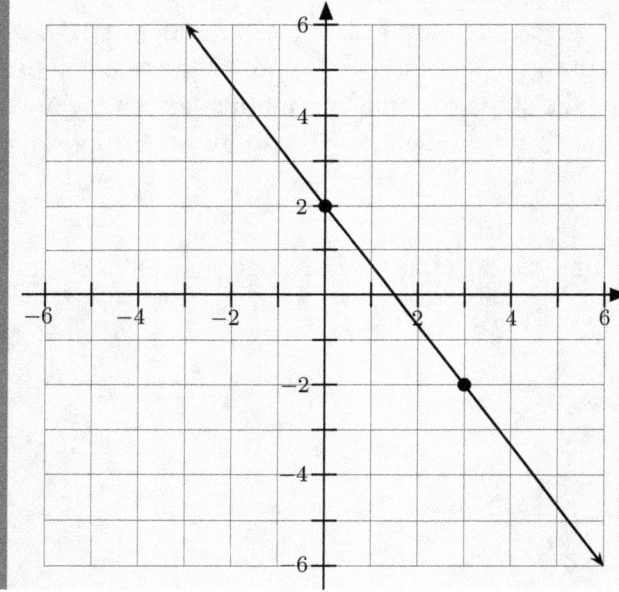

2.5 Slope-Intercept Form   155

Exercise 8 **You Try**
Find the equation of the line using the given graph. Write your answer in slope-intercept form.

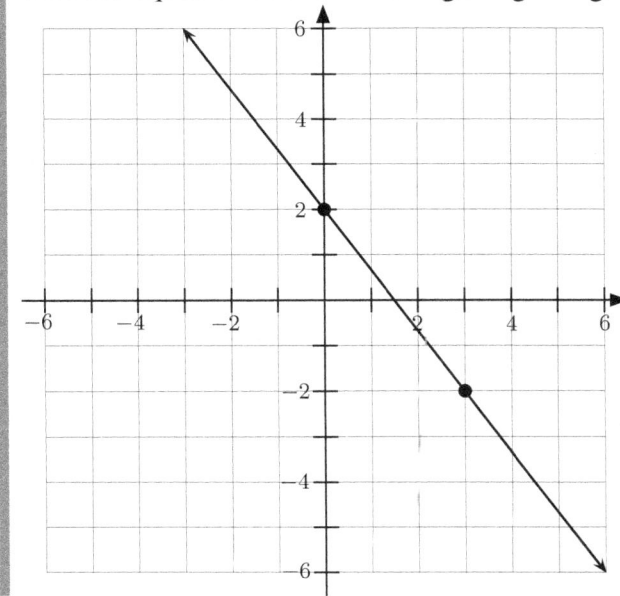

Equations of Horizontal and Vertical Lines

Lines, that are neither horizontal nor vertical, will always have a y-intercept and slope. Its equation will always be in two variables. On the other hand, horizontal and vertical lines are *special* in that their equation will only be in terms of one variable. Pay special attention to the slope and y-intercept, if any.

Example 7 Find the equation of the line using the given graph.

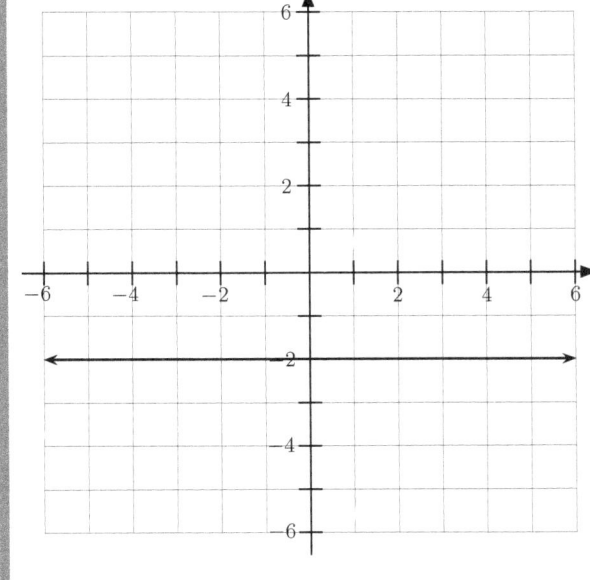

Solution.
We can read from the graph that the y-intercept is at the point, $(0, -2)$. That means, $b = -2$. Let's choose $(1, -2)$ as the second point on the graph. Using these two points in the slope formula we obtain

$$m = \frac{y_2 - y_1}{x_2 - x_1} \quad \text{Substitute values}$$

$$= \frac{-2 - (-2)}{1 - 0} \quad \text{Perform the operation}$$

$$= \frac{0}{1} \quad \text{Simplify fraction}$$

$$= 0 \quad \text{Slope of the line}$$

Notice that we have a line that is not steep at all. Therefore, the slope must be 0.

The equation of this horizontal line is $y = 0x + (-2)$ or simply $y = -2$.

Example 8 Find the equation of the line using the given graph.

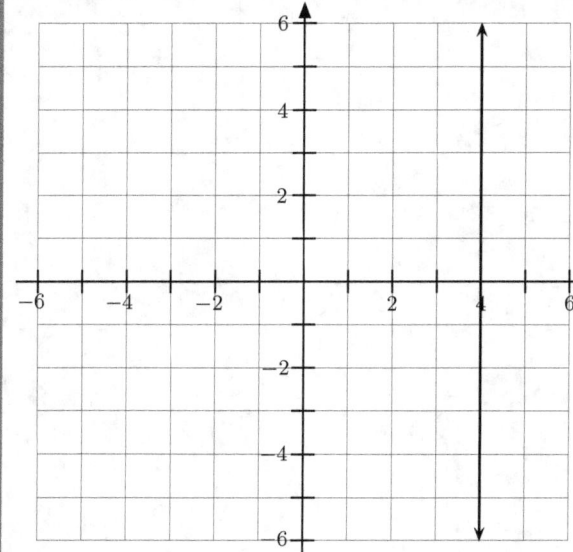

Solution.
Notice that the line does not intersect the y-axis, so there is no y-intercept. Hence, there is no b. Let us choose $(4, -2)$ and $(4, 1)$ as the two points on the graph. Using those two points in the slope formula, we obtain

$$m = \frac{y_2 - y_1}{x_2 - x_1}$$ Substitute values

$$= \frac{1 - (-2)}{4 - 4}$$ Perform the operation

$$= \frac{3}{0}$$ Slope is undefined

How do we find the equation of a vertical line when there is no b and the slope is undefined? Notice that any point on this vertical line will have 4 as the x-value, no matter what the y-value is. That is, any point on the line will have the form $(4, y)$.

The equation of this line is $x = 4$.

Exercise 9 Class Example
Find the equation of the line using the given graph.
a)  b)

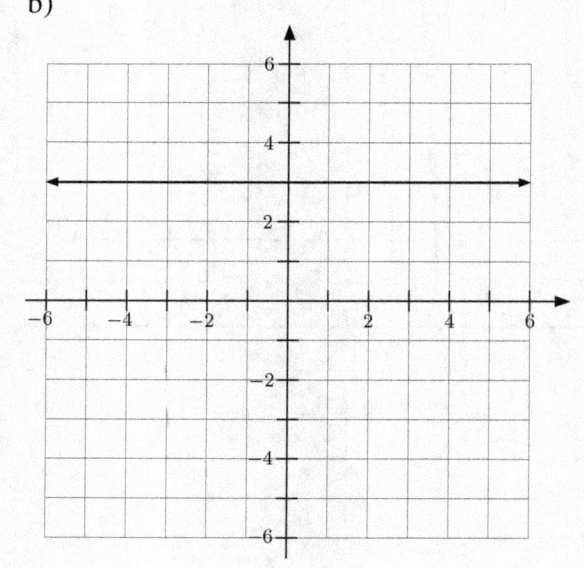

2.5 Slope-Intercept Form

Exercise 10 You Try

Find the equation of the line using the given graph.

a)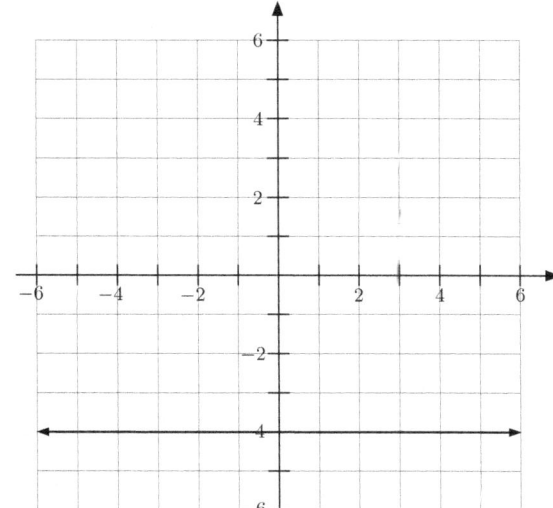

b)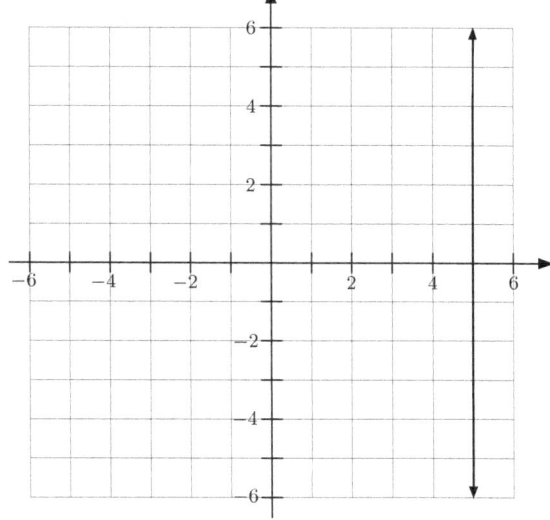

Applications

Let us take a look at some application problems involving lines. We will find the slope and the y-intercept and use that information to find the equation of the line.

Example 9 A moving truck rental company charges a $20.00 initial fee plus $0.90 per mile driven.

a) Write an equation that relates the number of miles driven, x, with the total cost of renting a truck, y.

b) How much will it cost if the truck went 35 miles?

c) If you were charged $66.80, how many miles was the truck driven?

Solution.

a) Let us first start by making a table of values where x is the number of miles driven and y is the total cost of renting a truck with this company.

x (miles driven)	y (Truck rental cost)
0	$20.00
1	$20.90
2	$21.80
3	$22.70
4	$23.60
5	$24.50

From the table, we see that the y-intercept, which is also the initial fee, is $(0, 20.00)$, making $b = 20.00$. From the problem, we see that for every mile driven, an additional $0.90 is added to the truck rental cost. This is the slope, $m = 0.90$.

Therefore, the equation of the line is $y = 0.90x + 20.00$.

b) To find the cost of renting the truck for 35 miles, substitute $x = 35$ and solve for y.

$$y = 0.90x + 20.00 \quad \text{Substitute } x = 35$$
$$= 0.90(35) + 20.00 \quad \text{Perform indicated operation}$$
$$= 51.50 \quad \text{Our Solution}$$

The truck rental will cost $51.50 to go 35 miles.

c) To find how many miles was the truck driven if the rental cost $66.80, substitute $y = 66.80$ and solve for x.

$$66.80 = 0.90x + 20.00 \quad \text{Subtract 20.00 from each side}$$
$$46.80 = 0.90x \quad \text{Divide each side by 0.90}$$
$$= 52 \quad \text{Our Solution}$$

The truck was driven for 52 miles if the rental cost $66.80.

Exercise 11 **Class Example**
In Seattle, a taxi charges a $2.60 meter drop charge (applied as soon as you enter the taxi) plus a distance charge of $2.70 per mile.
  a) Write an equation that relates the amount of miles for a taxi ride, x, with the total cost of the ride, y.

2.5 Slope-Intercept Form

b) How much will the taxi ride cost to go 25 miles?

c) If you paid $44.30, how many miles did you go?

Exercise 12 **You Try**
A student club wants to sell t-shits to generate funds for their expenses. A print-shop charges $30 for the t-shirt design plus $7.50 to print each shirt.
  a) Write an equation that relates the number of t-shirts printed, x, with the total cost of printing them, y.

  b) How much will it cost to print 250 t-shirts?

  c) How many t-shirts were printed if it cost the student club $967.50?

Exercise 13 You Try

According to a government report, the fuel economy of new U.S. cars and trucks hit a record 24.7 miles per gallon in the 2016 model year.
(Source: www.reuters.com)

a) Write an equation that relates the number of gallons consumed, x, with the distance the vehicle travels, y, in miles.

b) How many miles will new cars go if it consumes 5 gallons?

c) If a new car went 172.9 miles, how many gallons of gas did it consume?

2.5: Exercises

Write the slope-intercept form of the equation of each line, given the slope and the y-intercept.

1. Slope $= 2$, y-intercept $= (0, 5)$
2. Slope $= -6$, y-intercept $= (0, 4)$
3. Slope $= 1$, y-intercept $= (0, -4)$
4. Slope $= -1$, y-intercept $= (0, -2)$
5. Slope $= -\frac{3}{4}$, y-intercept $= (0, -1)$
6. Slope $= -\frac{1}{4}$, y-intercept $= (0, 3)$
7. Slope $= \frac{1}{3}$, y-intercept $= (0, 1)$
8. Slope $= \frac{2}{5}$, y-intercept $= (0, 5)$

Write the slope-intercept form of the equation of the line whose graph is shown.

9.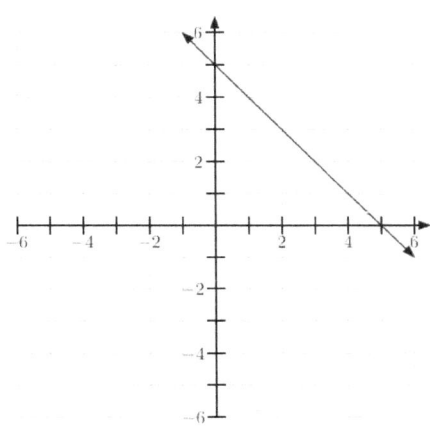

11.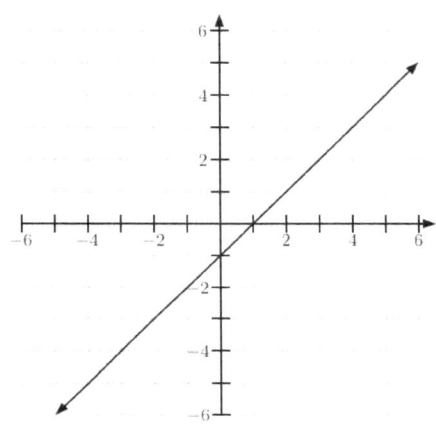

10.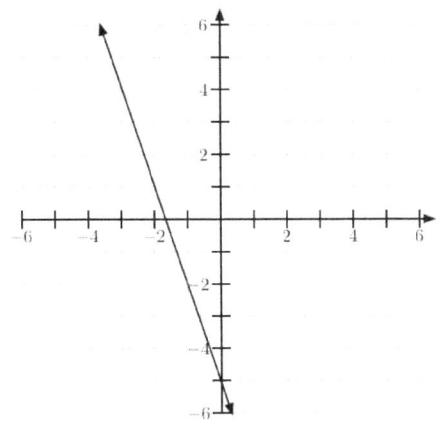

12.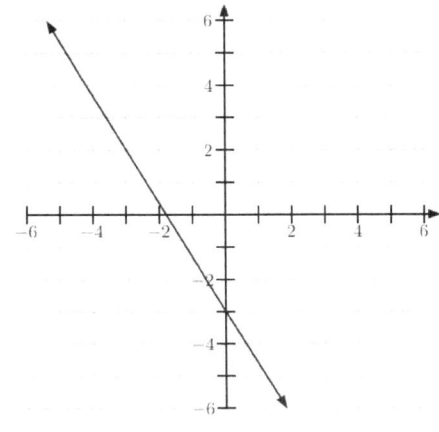

13.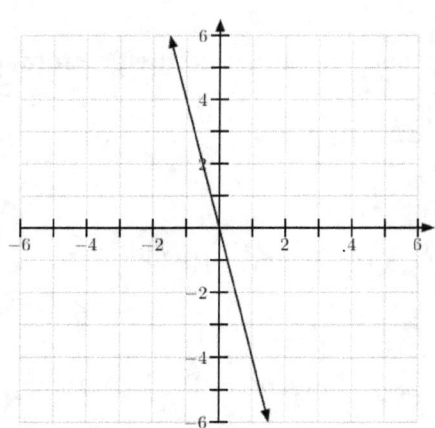

14.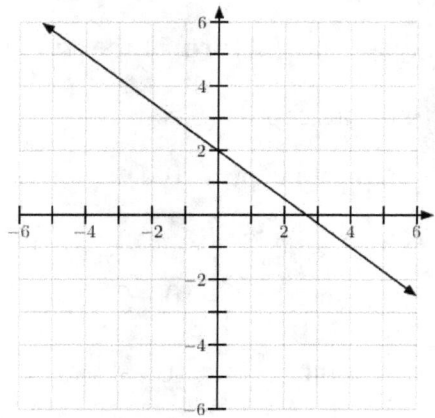

Write the slope-intercept form for the given linear equation.

15. $x + 10y = -37$

16. $x - 10y = 3$

17. $2x + y = -1$

18. $6x - 11y = -70$

19. $7x - 3y = 24$

20. $4x + 7y = 28$

21. $x = -8$

22. $x - 7y = -42$

23. $y - 4 = -(x + 5)$

24. $y - 5 = \dfrac{5}{2}(x - 2)$

25. $y - 4 = 4(x - 1)$

26. $y - 3 = -\dfrac{2}{3}(x + 3)$

27. $y + 5 = -4(x - 2)$

28. $0 = x - 4$

29. $y + 1 = -\dfrac{1}{2}(x - 4)$

30. $y + 2 = \dfrac{6}{5}(x + 5)$

2.5 Slope-Intercept Form

Sketch the graph of each line.

31. $y = \dfrac{1}{3}x + 4$

32. $y = -\dfrac{x}{5} - 4$

33. $y = \dfrac{6}{5}x - 5$

34. $y = -\dfrac{3}{2}x - 1$

35. $2y = 3x$

36. $x + 5y + 20 = 0$

37. $x - y + 3 = 0$

38. $4x + 5 = 5y$

39. $y + 4 - 3x = 0$

40. $2y - 8 = 6x$

41. $-3y = -5x + 9$

42. $x + y = 0$

Applications.

43. Elmer, who weighs 250 lbs, joins a diet program that promises to help him lose 2 pounds each week or his money back. Assume Elmer's weight loss is linear.

    (a) Write a linear equation in slope-intercept form, that represents the given situation. Let y represent Elmer's weight after being on the diet program for x weeks.

    (b) How much does Elmer weigh after 2.5 weeks if he follows the diet program?

    (c) If Elmer sticks with the program, how many weeks will it take before he gets to his target weight of 175 lbs?

44. Leigh is a heavy smoker, smoking 30 cigarettes a day, but he is determined to quit smoking. He is in a new program that will help him quit. The goal is to cut back 3 cigarettes every 2 weeks. Let x be the number of weeks Leigh has been in the program and let y be the number of cigarettes smoked per week. Assume the cigarettes smoked per week follows a linear pattern.

    (a) Write a linear equation, in slope-intercept form, that represents the given situation.

    (b) Identify the y-intercept.

    (c) Interpret the meaning of the y-intercept in the context of the given situation.

    (d) If Leigh follows the program, how many weeks will it take before he quits smoking entirely?

2.6 Point-Slope Form

Objective: To use the point-slope form to write the equation of a line

If we know the slope of a line and we know the y-intercept, we can easily write the equation of the line However, there are many occasions when we do not know the y-intercept but we know a different point on the line. If this is the case, we have another useful formula for finding the equation of the line.

If we let the slope of an equation be m, a specific point on the line be (x_1, y_1), and any other point on the line be (x, y), we can use the slope formula to obtain the following.

$$\frac{y - y_1}{x - x_1} = m \qquad \text{Multiply each side by } (x - x_1)$$

$$(x - x_1)\frac{y - y_1}{x - x_1} = m(x - x_1) \qquad \text{Simplify}$$

$$y - y_1 = m(x - x_1) \qquad \text{Point-Slope Form}$$

Point-Slope Form

Given a slope, m and a point, (x_1, y_1), the equation of the line in point-slope form is

$$y - y_1 = m(x - x_1)$$

Example 1 Find the equation of the line through the point $(1, 4)$ with a slope of -3.

Solution.

$$y - y_1 = m(x - x_1) \qquad \text{Substitute } m = -3 \text{ and } (x_1, y_1) = (1, -4)$$
$$y - 4 = -3(x - 1) \qquad \text{Equation of the line in point-slope form}$$

It is often preferred that final answers be written in slope-intercept form, $y = mx + b$. If the direction asks for the answer in slope-intercept form, distribute the slope and then solve for y.

Example 2 Find the equation of the line through the point $(6, -2)$ with a slope of $-\frac{2}{3}$. Write the answer in slope-intercept form.

2.6 Point-Slope Form

Solution.

$$y - y_1 = m(x - x_1)$$ Substitute $m = -\frac{2}{3}$ and $(x_1, y_1) = (6, -2)$

$$y - (-2) = -\frac{2}{3}(x - 6)$$ Simplify signs and distribute slope

$$y + 2 = -\frac{2}{3}x + 4$$ Subtract 2 from each side

$$y = -\frac{2}{3}x + 2$$ Equation of the line in slope-intercept form

Exercise 1 Class Example

Find the equation of the line through the given point and with the given slope. Write the answer in slope-intercept form.

a) line goes through the point $(4, -3)$ with slope 2

b) line goes through the point $(-1, 2)$ with slope $-\frac{3}{2}$

c) line goes through the point $(5, 8)$ with an undefined slope

Exercise 2 You Try
Find the equation of the line through the given point and with the given slope. Write the answer in slope-intercept form.

a) line goes through the point $(1, 1)$ with slope 6

b) line goes through the point $(-4, -5)$ with slope $-\dfrac{3}{4}$

c) line goes through the point $(0, 7)$ with a slope 0

Finding the Equation of a Line given Two Points

If we are given two points but not the slope, we can still find the equation of the line. First find the slope using the two given points. Then, with the slope and one of the given points, substitute the information into the point-slope form to find the equation of the line.

Example 3 Find the equation of the line through the points $(-2, 5)$ and $(4, -1)$. Write the answer in slope-intercept form.

Solution.

2.6 Point-Slope Form

Let $(x_1, y_1) = (-2, 5)$ and $(x_2, y_2) = (4, -1)$. Find the slope.

$$m = \frac{y_2 - y_1}{x_2 - x_1} \qquad \text{Substitute values}$$

$$m = \frac{-1 - 5}{4 - (-2)} \qquad \text{Perform indicated operation}$$

$$m = \frac{-6}{6} \qquad \text{Simplify}$$

$$m = -1 \qquad \text{Our slope}$$

Substitute the slope and one of the points into the point-slope form.

$$y - y_1 = m(x - x_1) \qquad \text{Use the point } (-2, 5)$$
$$y - 5 = -1(x - (-2)) \qquad \text{Simplify signs}$$
$$y - 5 = -1(x + 2) \qquad \text{Distribute slope}$$
$$y - 5 = -x - 2 \qquad \text{Add 5 to each side}$$
$$y = -x + 3 \qquad \text{Equation of line in slope-intercept form}$$

Example 4 Find the equation of the line through the points $(1, -1)$ and $(-3, -2)$. Write the answer in slope-intercept form.

Solution.
Let $(x_1, y_1) = (1, -1)$ and $(x_2, y_2) = (-3, -2)$. Find the slope.

$$m = \frac{y_2 - y_1}{x_2 - x_1} \qquad \text{Substitute values}$$

$$m = \frac{-2 - (-1}{-3 - 1)} \qquad \text{Perform indicated operation}$$

$$m = \frac{-1}{-4} \qquad \text{Simplify}$$

$$m = \frac{1}{4} \qquad \text{Our slope}$$

Substitute the slope and one of the points into the point-slope form.

$$y - y_1 = m(x - x_1)$$ Substitute value

$$y - (-1) = \frac{1}{4}(x - 1)$$ Simplify signs and distribute slope

$$y + 1 = \frac{1}{4}x - \frac{1}{4}$$ Subtract 1 from each side

$$y = \frac{1}{4}x - \frac{5}{4}$$ Equation of line in slope-intercept form

Exercise 3 Class Example
Find the equation of the that goes through the following points. Write the answer in slope-intercept form.

a) $(1, -3)$ and $(3, -7)$

b) x-intercept, 5 and y-intercept, -2

Exercise 4 You Try
Find the equation of the that goes through the following points. Write the answer in slope-intercept form.

a) $(0, 2)$ and $(4, 6)$

b) $(-5, 5)$ and $(1, 2)$

2.6 Point-Slope Form

If we are given the graph of a line, we can figure out its equation by locating 2 points on the line. Using those 2 points, first find the slope and then substitute it into the point-slope form.

Example 5 Find the equation of the line shown. Write the answer in slope-intercept form.

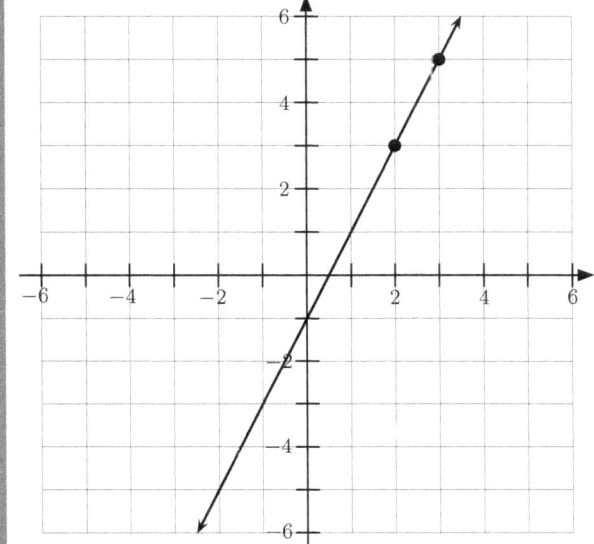

Solution.
Pick 2 points on the line. In this case, we will use $(x_1, y_1) = (2, 3)$ and $(x_2, y_2) = (3, 5)$. Now we find the slope

$$m = \frac{y_2 - y_1}{x_2 - x_1} \quad \text{Substitute values}$$

$$m = \frac{5 - 3}{3 - 2} \quad \text{Perform operation}$$

$$m = \frac{2}{1} \quad \text{Simplify}$$

$$m = 2 \quad \text{Our slope}$$

Substitute the slope and one of the points into the point-slope form.

$$y - y_1 = m(x - x_1) \quad \text{Substitute values}$$

$$y - 3 = 2(x - 2) \quad \text{Distribute slope}$$

$$y - 3 = 2x - 4 \quad \text{Add 3 to each side}$$

$$y = 2x - 1 \quad \text{Equation of line in slope-intercept form}$$

Exercise 5 **Class Example**
Find the equation of the line shown. Write the answer in slope-intercept form, if possible.

a)

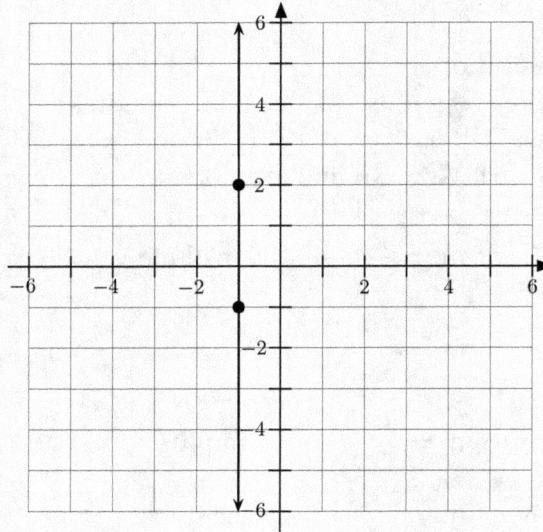

b)

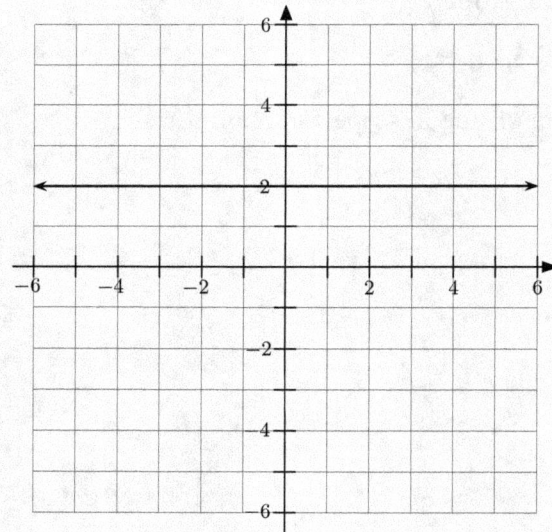

2.6 Point-Slope Form

Exercise 6 You Try
Find the equation of the line shown. Write the answer in slope-intercept form, if possible.

a)

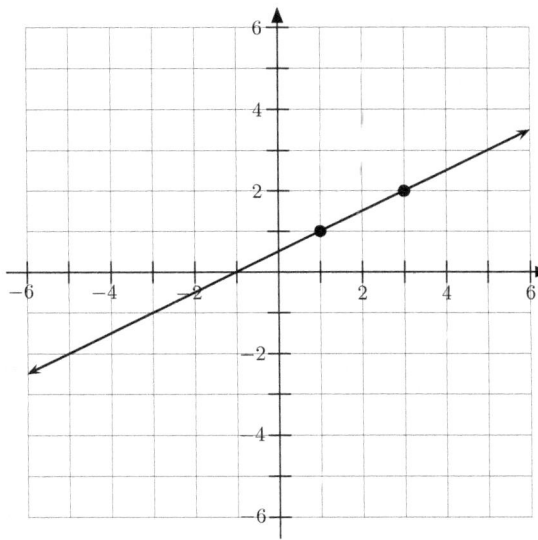

b)

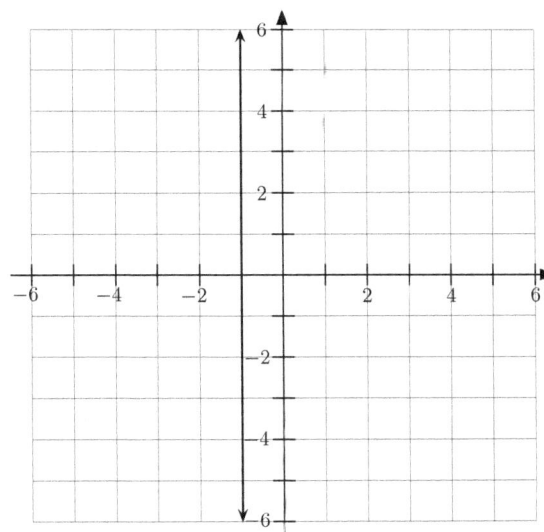

Applications

For the application problems in this section, we will identify two points and use those two points to calculate the slope. Then, we will use the slope and one of the points to determine the equation of the line.

Example 6 If you are between 120-140 lbs and walk at a rate of 3 mph, in 15 minutes, you would burn 50 calories and in 30 minutes, 100 calories. (Source: Good Housekeeping)
 a) Let x be the walking time and y be the calories burned. Find two ordered pairs that represent the given information.
 b) Identify the slope and interpret the slope in the context of the given situation.
 c) Write a linear equation, in slope-intercept form, that represents the given situation.
 d) You weigh 130 lbs. How many calories will you burn if you walk for 40 minutes at 3 mph?
 e) If you want to burn 125 calories, how long do you have to walk at 3 mph if you weigh 130 lbs?

Solution.

a) Let us first find two ordered pairs of the form (walking time, calories burned). We have (15 minutes, 50 calories) and (30 minutes, 100 calories).

b) Let $(x_1, y_1) = (15, 50)$ and $(30, 100)$. Find the slope using the slope formula.

$$m = \frac{y_2 - y_1}{x_2 - x_1} \qquad \text{Substitute values}$$

$$= \frac{100 - 50}{30 - 15} \qquad \text{Perform indicated operation}$$

$$= \frac{50}{15} \qquad \text{Simplify fraction}$$

$$= \frac{10}{3} \approx 3.3 \qquad \text{Our slope}$$

The slope $\frac{10}{3}$ means that if you are between 120-140 lbs and walk at a rate of 3 mph, you will burn 10 calories for every 3 minutes walked.
The slope 3.3 means that if you are between 120-140 lbs and walk at a rate of 3 mph, you will burn 3.3 calories for every minute walked.

c) To find the equation of the line, substitute the slope and one of the points into the point-slope

2.6 Point-Slope Form

form.

$$y - y_1 = m(x - x_1)$$ Substitute values

$$y - 50 = \frac{10}{3}(x - 15)$$ Distribute slope

$$y - 50 = \frac{10}{3}x - 50$$ Add 50 to each side

$$y = \frac{10}{3}x$$ Equation of the line in slope-intercept form

d) To figure out how many calories are burned in 40 minutes, substitute $x = 40$ and solve for y.

$$y = \frac{10}{3}(40)$$ Simplify fraction

$$= \frac{400}{3} \approx 133.3$$ Our solution rounded to the nearest tenth

If you are 130 lbs and walk for 40 minutes at a rate of 3 mph, you will burn approximately 133.3 calories.

e) To figure out how long to walk in order to burn 125 calories, substitute $y = 125$ and solve for x.

$$125 = \frac{10}{3}x$$ Multiply each side by $\frac{3}{10}$

$$\frac{3}{10}(125) = x$$ Perform indicated operation

$$\frac{375}{10} = 37.5 = x$$ Our Solution

If you are 130 lbs and walk at a rate of 3 mph, in order to burn 125 calories, you will need to walk for 37.5 minutes.

Exercise 7 Class Example

The World Health Organization (WHO) has been monitoring the percent of population without improved drinking water source. Five years after they began their study, 20% of the world population were without improved drinking water source. Ten years after they began their study, 17% of the world population were without improved drinking water source. Let y be the percent of world population without improved drinking water source and let x be the years after WHO began their study. (Source: WHO/UNICEF Joint Monitoring Programme for Water Supply and Sanitation)

a) Find two ordered pairs that represent the given information.

b) Identify the slope and interpret the slope in the context of the given situation.

c) Write a linear equation, in slope-intercept form, that represents the given situation.

d) What is the y-intercept? Interpret the meaning of the y-intercept in the context of the situation.

e) If the trend continues, what percent of the world population would you expect to be without improved drinking water source 12 years after WHO began their study?

f) Approximately how many years after the study began did 14% of the world population not have improved drinking water source?

2.6 Point-Slope Form

Exercise 8 You Try

Tuition for Washington State residents taking classes at North Seattle College during the academic year 2017-2018 increases at the same rate for the first 10 credits. If you take 3 credits, tuition is $322.77. If you take 5 credits, tuition is $537.95. Let y represent tuition for x credits taken.

a) Find two ordered pairs that represent the given information.

b) Identify the slope and interpret the slope in the context of the given situation.

c) Write a linear equation, in slope-intercept form, that represents the given situation.

d) Suppose a Washington State resident wants to take 10 credits, how much will the tuition be?

e) If a Washington State resident paid $860.72 in tuition, how many credits is he or she taking?

2.6: Exercises

Write the point-slope form of the equation of the line with the given properties.

1. through $(-1,-5)$, slope = 9
2. through $(2,2)$, slope = $\frac{1}{2}$
3. through $(0,-7)$, slope = $-\frac{1}{4}$
4. through $(-4,1)$, slope = 0
5. through $(2,3)$, slope is undefined
6. through $(-1,0)$, slope = $-\frac{5}{4}$
7. through $(-4,3)$ and $(-3,1)$
8. through $(1,6)$ and $(-3,6)$
9. through $(-1,-4)$ and $(-5,0)$
10. through $(-8,1)$ and $(-8,4)$
11. through $(-4,-2)$ and $(0,4)$
12. through $(3,5)$ and $(-5,3)$

Write the slope-intercept form of the equation of the line with the given properties.

13. through $(2,-2)$, slope = 1
14. thorugh $(-1,-7)$, slope = 2
15. through $(3,4)$, undefined slope
16. through $(4,0)$, slope = $-\frac{3}{2}$
17. through $\left(-\frac{1}{2},-\frac{3}{4}\right)$, slope = 0
18. through $(-2,-2)$, slope = $-\frac{2}{3}$
19. through $(0,2)$ and $(5,-3)$
20. through $(0,1)$ and $(-3,0)$
21. through $(4,1)$ and $(1,4)$
22. through $\left(\frac{2}{7},-1\right)$ and $\left(\frac{2}{7},-2\right)$
23. through $(-5,1)$ and $(-1,-2)$
24. through $(1,-1)$ and $(-5,-4)$

Write the slope-intercept form of the equation of the line shown.

25.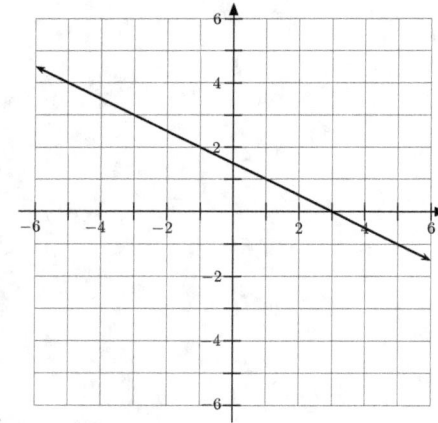

26.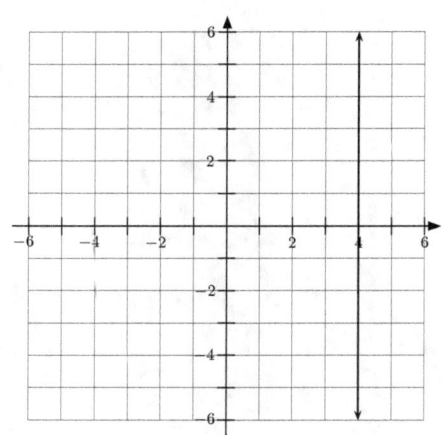

2.6 Point-Slope Form

27.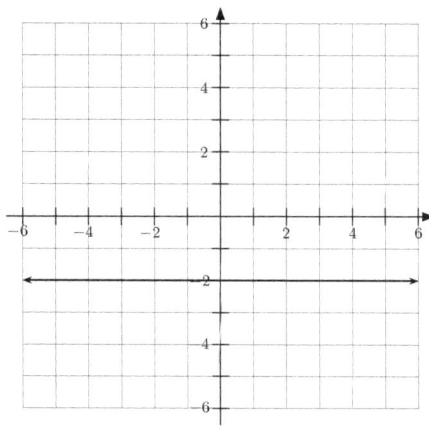
28.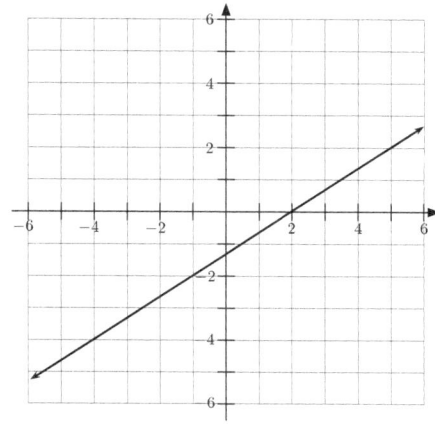

Applications.

29. A yogurt company recently released a new yogurt blend. The company expects to make $68,000 in profit on the new brand at the end of the first year. At the end of the 4th year, the company expects profit to be $272,000. Let y represent profit at the end of x years.

    (a) Write the ordered pairs that represent the above information.

    (b) Identify the slope and interpret the slope in the context of the situation.

    (c) Write a linear equation in slope-intercept form that represents the relationship between year and profit.

    (d) Use the equation to find the company's profit at the end of the 6th year.

30. For the academic year 2015-2016, tuition for International students taking classes towards a BAS degree at North Seattle College increases at the same rate for the first 10 credits. If you take 2 credits, tuition is $1,197.68. If you take 7 credits, tuition is $4,191.88. Let y represent tuition for x credits taken.

    (a) Write the ordered pairs that represent the above information.

    (b) Identify the slope and interpret the slope in the context of the situation.

    (c) Write a linear equation in slope-intercept form that represents the relationship between the number of credits and tuition.

    (d) Suppose an international student wants to take 10 credits of classes. How much will the tuition be?

31. Cars usually depreciate linearly. You bought a Kia Soul. A year later, the value of your Kia is $17,530. Three years later, the value of your Kia is $11,849. Let y be the value of your Kia x years after you purchased your car.

    (a) Write the ordered pairs that represent the above information.

    (b) Identify the slope.

    (c) Interpret the slope in the context of the situation.

(d) Write a linear equation in slope-intercept form that represents the relationship between the number of years you own the Kia and its value.

(e) What is the y-intercept?

(f) Interpret the meaning of the y-intercept in the context of the situation.

2.7 Parallel and Perpendicular Lines

Objective: To find the equation of a line that is parallel or perpendicular to another line

Slopes of Parallel and Perpendicular Lines

Exercise 1 You Try

Each number shows a pair of parallel lines.

a) Find the slope of each line.

b) What do you discover?

(i)

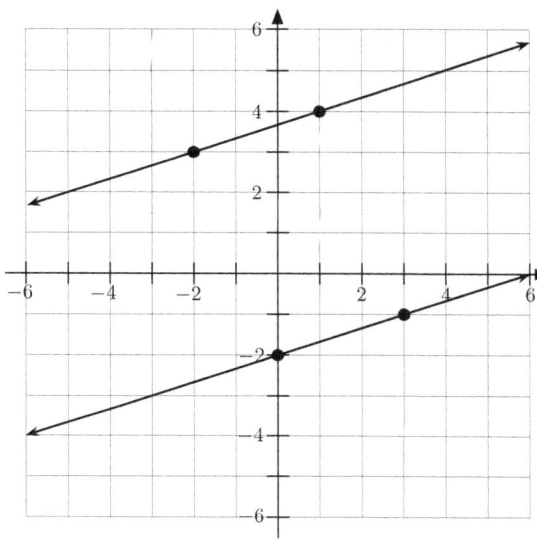

(ii)

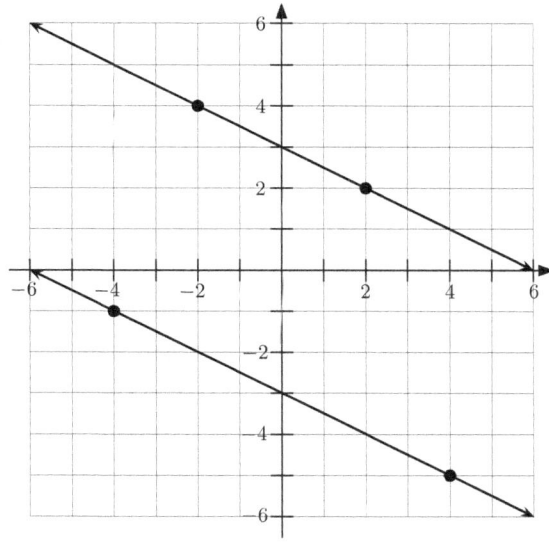

Exercise 2 **You Try**
Each number shows a pair of perpendicular lines.

a) Find the slope of each line.   b) What do you discover?

(i)

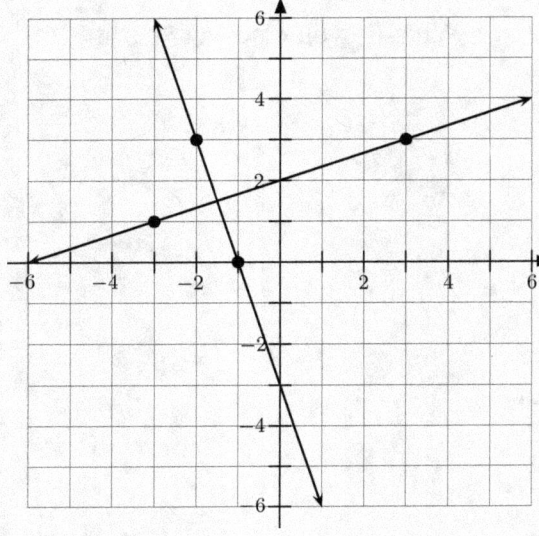

(ii)

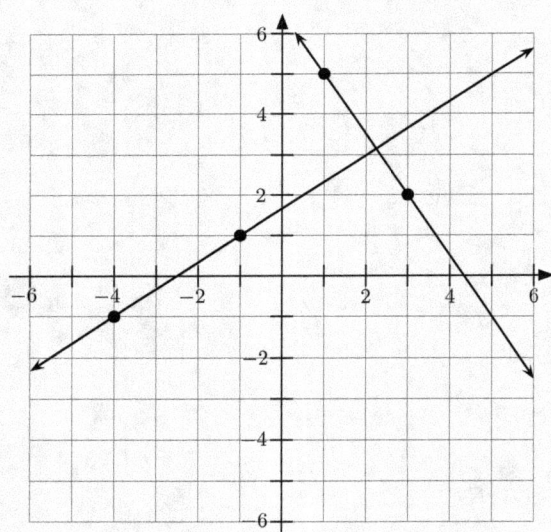

From the graphs above, we see that parallel lines have the same slope while perpendicular lines have slopes that are negative reciprocals of one another.

2.7 Parallel and Perpendicular Lines

Slopes of Parallel and Perpendicular Lines

Suppose one line has slope, m_1, and the another line has slope, m_2.

- If the two lines are parallel to each other, then their slopes are the same. That is, $m_1 = m_2$.
- If the two lines are perpendicular to each other, then one slope is the negative reciprocal of the other. That is, $m_1 = -\dfrac{1}{m_2}$.

Example 1 Given the line $y = \dfrac{3}{5}x - 7$.

a) Find the slope of a line parallel to the given line.

b) Find the slope of a line perpendicular to the given line.

Solution.
We see that the line $y = \dfrac{3}{5}x - 7$ has a slope of $\dfrac{3}{5}$.

a) A line parallel to the given line has the same slope as the given line. Therefore, the slope of the parallel line is $\dfrac{3}{5}$.

b) A line perpendicular to the given line has a slope that is the negative reciprocal of the given line's slope. Therefore, the perpendicular line has slope $-\dfrac{5}{3}$.

Exercise 3 Class Example Given the line $y = \dfrac{1}{2}x$.

a) Find the slope of a line parallel to the given line.

b) Find the slope of a line perpendicular to the given line.

Exercise 4 **You Try** Given the line $y = -\dfrac{5}{8}x - 1$.

a) Find the slope of a line parallel to the given line.

b) Find the slope of a line perpendicular to the given line.

Example 2 Given the line $2x - y = 3$.

a) Find the slope of a line parallel to the given line.

b) Find the slope of a line perpendicular to the given line.

Solution.
First, we need to rewrite the equation of the line in slope-intercept form so we can easily read its slope.

$$2x - y = 3$$
$$\underline{-2x = -2x} \qquad \text{Subtract 2x from each side}$$
$$-y = -2x + 3$$

$$\dfrac{-y}{-1} = \dfrac{-2x}{-1} + \dfrac{3}{-1} \qquad \text{Divide each term by } -1$$

$$y = 2x - 3 \qquad \text{Equation of the line in slope-intercept form}$$

We see that the slope of the line is 2 or $\dfrac{2}{1}$, in fraction form.

a) A line parallel to the given line has the same slope as the given line. Therefore, the slope of the parallel line is 2.

b) A line perpendicular to the given line has a slope that is the negative reciprocal of the given line's slope. Therefore, the perpendicular line has slope $-\dfrac{1}{2}$.

2.7 Parallel and Perpendicular Lines

Example 3 Given the line $5y + 2x = 7$.

a) Find the slope of a line parallel to the given line.

b) Find the slope of a line perpendicular to the given line.

Solution.
First, we need to rewrite the equation of the line in slope-intercept form so we can easily read its slope.

$$5y + 2x = 7$$
$$\underline{-2x = -2x} \qquad \text{Subtract } 2x \text{ from each side}$$
$$5y = -2x + 7$$

$$\frac{5y}{5} = \frac{-2x}{5} + \frac{7}{5} \qquad \text{Divide each term by 5}$$

$$y = -\frac{2}{5}x + \frac{7}{5} \qquad \text{Equation of the line in slope-intercept form}$$

We see that the slope of the line is $-\frac{2}{5}$.

a) A line parallel to the given line has the same slope as the given line. Therefore, the slope of the parallel line is $-\frac{2}{5}$.

b) A line perpendicular to the given line has a slope that is the negative reciprocal of the given line's slope. Therefore, the perpendicular line has slope $\frac{5}{2}$.

Exercise 5 **Class Example**
For each of the given lines, do the following.

- Find the slope of a line parallel to the given line.
- Find the slope of a line perpendicular to the given line.

a) $6x - 5y = 2$
b) $x + y = 10$

Exercise 6 You Try
For each of the given lines, do the following.

- Find the slope of a line parallel to the given line.
- Find the slope of a line perpendicular to the given line.

a) $3x + 4y = 8$ b) $x - 3y = 6$

Now that we have learned how to find the slope of a line parallel to or perpendicular to another line, we will use this information to find the equation of the line. Let us take a look at some examples.

Example 4 Find the equation of the line that goes through the point $(9, -5)$ and is parallel to the line $2x + 3y = 6$. Write the answer in slope-intercept form.

Solution.
First, find the slope of the line $2x + 3y = 6$ by rewriting the equation in slope-intercept form.

$$2x + 3y = 6$$
$$\underline{-2x \qquad = -2x} \qquad \text{Subtract 2x from each side}$$
$$3y = -2x + 6$$

$$\frac{3y}{3} = \frac{-2x}{3} + \frac{6}{3} \qquad \text{Divide each term by 3}$$

$$y = -\frac{2}{3}x + 2 \qquad \text{Equation of the given line}$$

We see that the slope of the given line is $-\frac{2}{3}$. The line parallel to the given line will also have a slope of $-\frac{2}{3}$. To find the equation of the parallel line going through the point $(9, -5)$, substitute

2.7 Parallel and Perpendicular Lines

the slope and the given point into the point-slope form.

$$y - y_1 = m(x - x_1)$$ Substitute values

$$y - (-5) = -\frac{2}{3}(x - 9)$$ Change sign and distribute slope

$$y + 5 = -\frac{2}{3}x + 6$$ Subtract 5 from each side

$$y = -\frac{2}{3}x + 1$$ Equation of the parallel line

Example 5 Find the equation of the line that goes through the point $(-4, 1)$ and is perpendicular to the line $y - 2x = 5$. Write the answer in slope-intercept form.

Solution.
First, find the slope of the line $y - 2x = 5$ by rewriting the equation in slope-intercept form.

$$y - 2x = 5$$
$$\underline{+2x = +2x}$$ Add 2x to each side
$$y = -2x + 5$$ Equation of the given line

We see that the slope of the given line is -2. The line perpendicular to the given line will have a slope that is the negative reciprocal of -2 or $\frac{1}{2}$. To find the equation of the perpendicular line going through the point $(-4, 1)$, substitute the slope and the given point into the point-slope form.

$$y - y_1 = m(x - x_1)$$ Substitute values

$$y - 1 = \frac{1}{2}(x - (-4))$$ Change sign

$$y - 1 = \frac{1}{2}(x + 4)$$ Distribute slope

$$y - 1 = \frac{1}{2}x + 2$$ Add 1 to each side

$$y = \frac{1}{2}x + 3$$ Equation of perpendicular line

Exercise 7 Class Example
Find the equation of the line with the information. Write the answer in slope-intercept form.

a) the line that goes through the point $(5,-2)$ and is parallel to the line $5y-3x=0$.

b) the line line that goes through the point $(-7,3)$ and is perpendicular to the line $x-2y=10$

Exercise 8 You Try
Find the equation of the line with the given information. Write the answer in slope-intercept form.

a) the line that goes through the point $(6,-1)$ and is parallel to the line $4x+3y=-9$

b) the line that goes through the point $(-1,-3)$ and is perpendicular to the line $x-y=0$

2.7 Parallel and Perpendicular Lines

Horizontal and Vertical Lines

Recall that a horizontal line has a slope of 0 and a vertical line has an undefined slope. If a line is parallel to a horizontal line, then its slope will also be 0. The equation of a horizontal line will always take the form, $y = $ a real number. To find the slope of the line perpendicular to a horizontal line, it makes no sense to find the negative reciprocal of 0. The line perpendicular to a horizontal line is a vertical line and a vertical line has an undefined slope. The equation of a vertical line will always take the form $x = $ a real number.

Example 6 Find the equation of the line given the following information.
 a) a line goes through the point $(3,4)$ and is parallel to the line $x = -2$
 b) a line goes through the point $(3,4)$ and is perpendicular to the line $x = -2$

Solution.

 a) The line $x = -2$ is a vertical line and vertical lines have undefined slope. The line parallel to $x = -2$ and going through the point $(3,4)$ will also have an undefined slope with equation $x = 3$.

 b) The line perpendicular to the vertical line, $x = -2$ is a horizontal line with a slope of 0. The horizontal line going through the point $(3,4)$ will have equation $y = 4$.

Exercise 9 **Class Example**
Find the equation of the line given the following information.
 a) a line goes through the point $(-5,1)$ and is parallel to the line $y = -4$

 b) a line goes through the point $(-5,1)$ and is perpendicular to the line $y = -4$

Exercise 10 **You Try**
Find the equation of the line given the following information.
 a) a line goes through the point $(4,-7)$ and is parallel to the line $x = 1$

 b) a line goes through the point $(4,-7)$ and is perpendicular to the line $x = 1$

c) a line goes through the point $(-2, 6)$ and is parallel to the line $y = 0$

d) a line goes through the point $(-2, 6)$ and is perpendicular to the line $y = 0$

2.7 Parallel and Perpendicular Lines

2.7: Exercises

Complete the table by finding the appropriate slopes.

	Given Line	Slope of Parallel Line	Slope of Perpendicular Line
1.	$y = 4x - 5$		
2.	$x = y + 4$		
3.	$3y = x + 6$		
4.	$x + 2y = 8$		

Write the slope-intercept form of the equation of the described line.

5. through $(3,4)$ and parallel to $y = \dfrac{9}{2}x - 5$

6. through $(1,-1)$ and parallel to $y = -\dfrac{3}{4}x + 3$

7. through $(-3,-5)$ and perpendicular to $x + 2y = -4$

8. through $(2,-3)$ and perpendicular to $-2x + 5y = -10$

9. through $(1,-5)$ and perpendicular to $-x + y = 1$

10. through $(1,4)$ and parallel to $y = \dfrac{7}{5}x + 4$

11. through $(-1,3)$ and parallel to $y = -3x - 1$

12. through $(-4,-1)$ and parallel to $y = -\dfrac{1}{2}x + 1$

13. through $(2,5)$ and parallel to $x = 0$

14. through $(1,-2)$ and perpendicular to $-x + 2y = 2$

15. through $(5,2)$ and perpendicular to $5x + y = -3$

16. through $(-8,3)$ and parallel to $x = 4$

17. through $(1,3)$ and parallel to $-x + y = 1$

18. through $(7,-3)$ and perpendicular to $x = -2$

19. through $(-2,5)$ and perpendicular to $y - 2x = 0$

20. through $(-3,-5)$ and perpendicular to $3x + 7y = 0$

Rescue Roody! Rescue Roody!

21. Roody has to find the equation of the line that goes through the point $(-5, 3)$ and is perpendicular to $y = 4$. He is stumped because the line $y = 4$ has a slope of 0 but the line perpendicular to it has an undefined slope. Help him figure out how to work this problem.

2.8 Linear Inequalities in Two Variables

Objective: To be able to graph the set of solutions to a linear inequality in two variables

In the previous chapter, we solved linear inequalities in one variable such as $2x - 1 < 3$. The set of solutions, $x < 2$, is the set of all real numbers that satisfy the inequality $2x - 1 < 3$. We can graph this set on the number line as follows.

In this section, we will solve linear inequalities in two variables such as $2x - y < 3$. Since there are two variables, we will now need to specify a solution in the form of an ordered pair whose co-ordinates satisfy the inequality. **A solution** to a linear inequality in two variables is an ordered pair whose coordinates satisfy the inequality after substitution. The **solution set** of a linear inequality in two variables will be the collection of **all ordered pairs** that satisfy the inequality.

Example 1 Are the following ordered pair solutions to the inequality $y > 2x - 3$?

a) $(2, 3)$ b) $(1, -6)$ c) $(-1, -5)$

Solution.
In each of the problems, we will substitute the coordinates of the given ordered pair into the inequality and determine whether they satisfy the inequality or not.

a) Substitute $(x, y) = (2, 3)$ into $y > 2x - 3$.

$$3 \stackrel{?}{>} 2(2) - 3$$
$$3 \stackrel{?}{>} 4 - 3$$
$$3 > 1 \ \checkmark$$

Since the ordered pair satisfies the inequality, $(2, 3)$ is a solution to $y > 2x - 3$.

b) Substitute $(x, y) = (1, -6)$ into $y > 2x - 3$.

$$-6 \stackrel{?}{>} 2(1) - 3$$
$$-6 \stackrel{?}{>} 2 - 3$$
$$-6 \not> -1$$

Since the ordered pair does not satisfy the inequality, $(-1, -5)$ is not a solution to $y > 2x - 3$.

c) Substitute $(x,y) = (-1,-5)$ into $y > 2x - 3$.

$$-5 \stackrel{?}{>} 2(-1) - 3$$
$$-5 \stackrel{?}{>} -2 - 3$$
$$-5 \not> -5$$

Since the ordered pair does not satisfy the inequality, $(-1,-5)$ is not a solution to $y > 2x - 3$.

Exercise 1 Class Example
Are the following ordered pair solutions to the inequality $2x - y \geq 4$?

a) $(3,1)$ \qquad b) $(0,-2)$ \qquad c) $\left(\dfrac{3}{4}, -\dfrac{5}{2}\right)$

Exercise 2 You Try
Are the following ordered pair solutions to the inequality $3x + 4y \leq 12$?

a) $(-1,5)$ \qquad b) $(0,0)$ \qquad c) $\left(-1, \dfrac{15}{4}\right)$

2.8 Linear Inequalities in Two Variables

Graphing Linear Inequalities in Two Variables

We now wish to describe the set of all ordered pair solutions to a linear inequality in two variables. Lets take a look at the graph of the line $y = 2x - 3$ and the locations of the ordered pairs $(2,3)$, $(1,-6)$ and $(-1,-5)$.

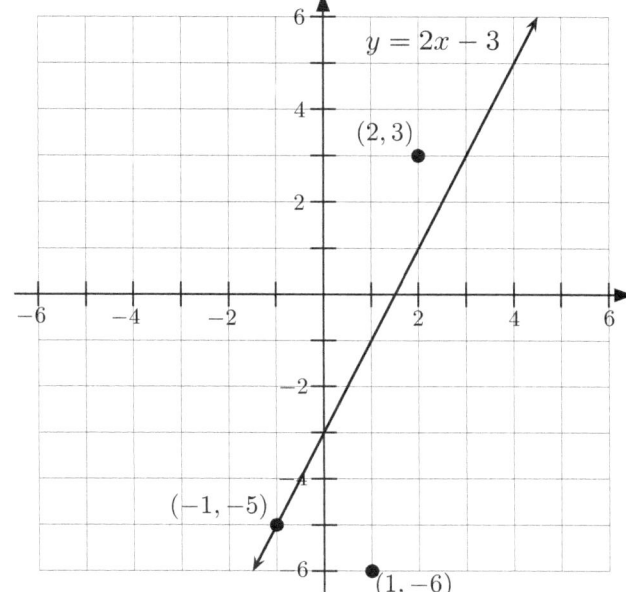

We note the following:

- The point $(-1,-5)$ is on the line since the point $(-1,-5)$ satisfies the equation, $y = 2x - 3$. In fact, all points on the line, $y = 2x - 3$ are precisely those whose coordinates satisfy the equation, $y = 2x - 3$.

- The point $(2,3)$ lies above the line, $y = 2x - 3$. Because of this, the point $(2,3)$ satisfies the inequality, $y > 2x - 3$. In fact, any point that lies above the line, $y = 2x - 3$, satisfies the inequality, $y > 2x - 3$, since its y-value is greater than the corresponding y-value of the point on the line directly below it.

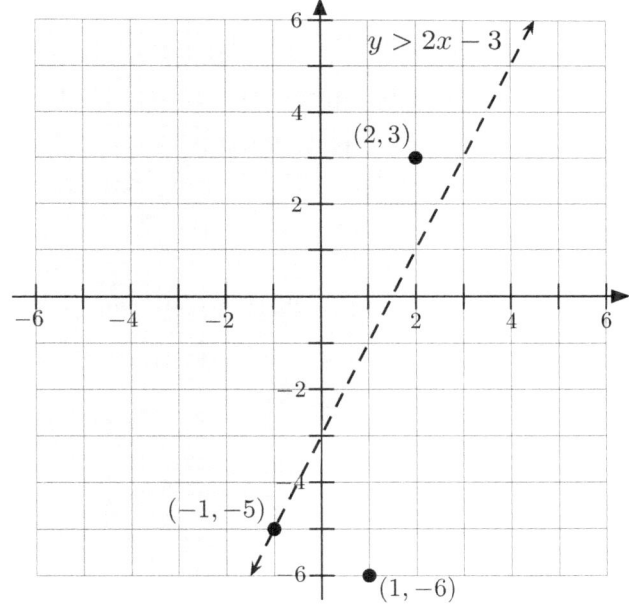

- The graph of the inequality, $y > 2x - 3$, will include all values of y that are greater than the values of y on the line, $y = 2x - 3$.

- This is shown by shading the region above the line. Since the inequality is a strict inequality, we do not include points that are on the line. This is denoted by drawing a dashed line on $y = 2x - 3$.

- We see that the point $(1,-6)$ lies below the line, $y = 2x - 3$. Because of this, the point $(1,-6)$ satisfies the inequality, $y < 2x - 3$.

In fact, any point that lies below the line, $y = 2x - 3$, satisfies the inequality, $y < 2x - 3$, since its y-value is less than the corresponding y-value of the point on the line directly above it.

The graph of the inequality, $y < 2x - 3$, will include all values of y that are less than the values of y on the line, $y = 2x - 3$.

This is shown by shading the region below the line.

Since the inequality is a strict inequality, we do not include the points that are on the line. This is denoted by drawing a dashed line on $y = 2x - 3$.

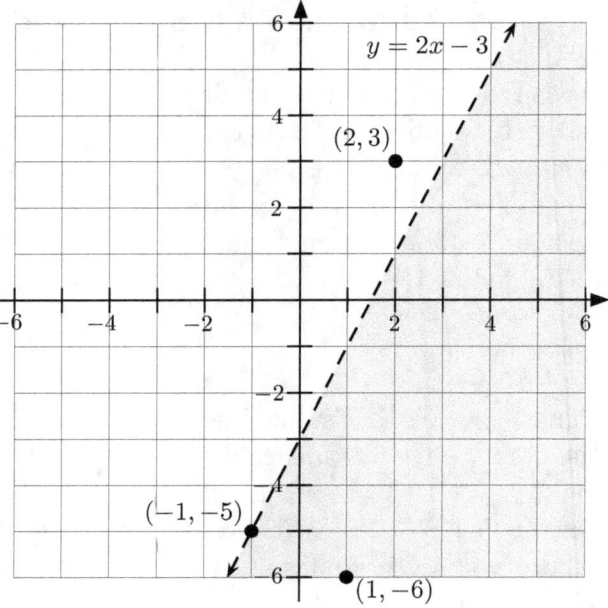

Graphing the Set of Solutions to a Linear Inequality in Two Variables:

Isolate y to the left side of the inequality. Then determine the slope and y-intercept.

- Graph the boundary line by replacing the linear inequality with a linear equality.

  ◇ Use a **dashed line** if the inequality is strictly less than or strictly greater than (that is, $<$ or $>$). This means that points on the line are not included in the solution set.

  ◇ Use a **solid line** if the inequality is not strict (that is, $\leqslant$ or $\geqslant$). This means that points on the line are included in the solution set.

- Shading

  ◇ Shade **above** the line if the inequality is greater than ($>$) or greater than or equal to ($\geqslant$).

  ◇ Shade **below** the line if the inequality is less than ($<$) or less than or equal to ($\leqslant$).

2.8 Linear Inequalities in Two Variables

Example 2 Graph the set of solutions to the following linear inequality.

a) $y \leq -\frac{1}{2}x + 4$

b) $3x - 4y < 12$

Solution.

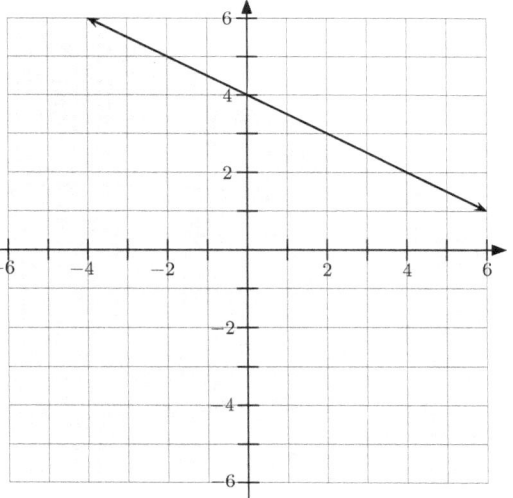

a) The variable y is already isolated to the left.

Next, determine the slope, $m = -\frac{1}{2}$ and y-intercept, $(0, 4)$.

Graph $y = -\frac{1}{2}x + 4$.

This is the boundary line. Since the inequality is non-strict ($\leq$), the boundary line will be a solid line.

We will shade below the boundary line because the inequality is less than or equal ($\leq$).

b) We will need to first solve for y. Don't forget to reverse the inequality when you multiply or divide by a negative number.

$$3x - 4y < 12$$
$$\underline{-3x \qquad -3x}$$ Subtract 3x from each side
$$-4y < -3x + 12$$

$$\frac{-4y}{-4} > \frac{-3x}{-4} + \frac{12}{-4}$$ Divide each side by -4

$$y > \frac{3}{4}x - 3$$ Inequality with y on the left

We can now determine the slope, $m = \frac{3}{4}$ and y-intercept, $(0, -3)$.

Graph $y = \frac{3}{4}x - 3$. This is the boundary line.

Since the inequality is strictly greater than $(>)$, the boundary line will be a dashed line.

We will shade above the boundary line because the inequality is greater than $(>)$.

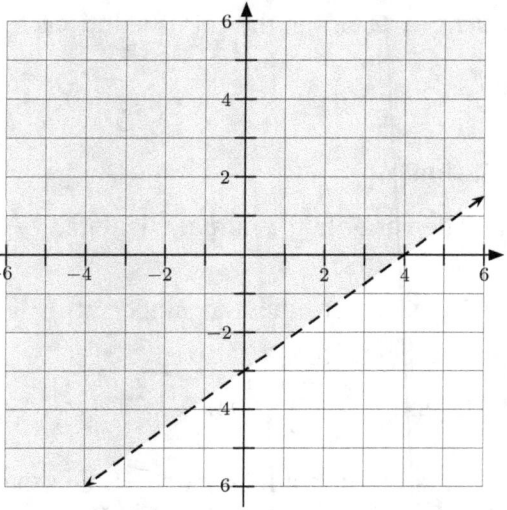

Exercise 3 Class Example
Graph the set of solutions to the following linear inequalities.

a) $4x + y < 7$

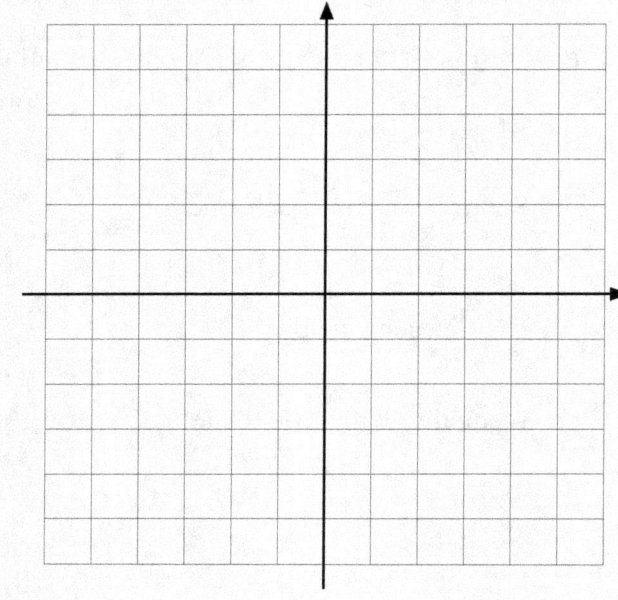

b) $x - y \leqslant 2$

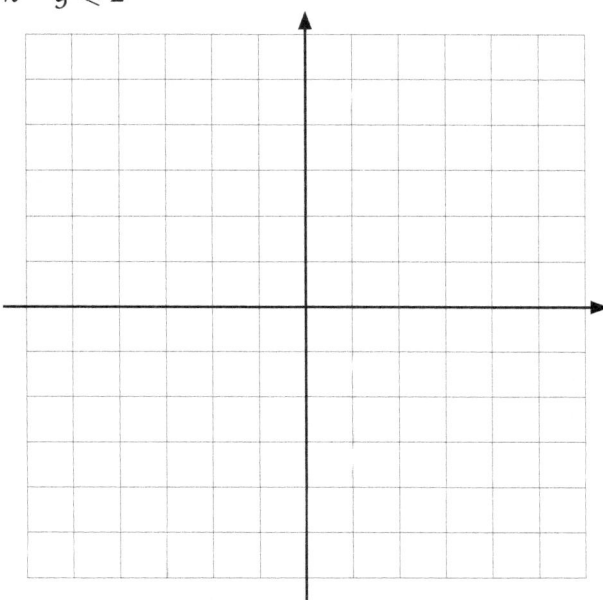

Exercise 4 You Try
Graph the set of solutions to the following linear inequality.

a) $y - 3x \geq 2$

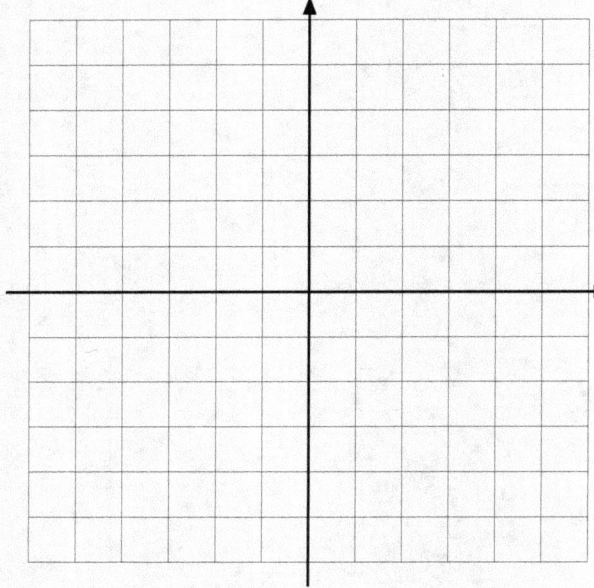

b) $x > y$

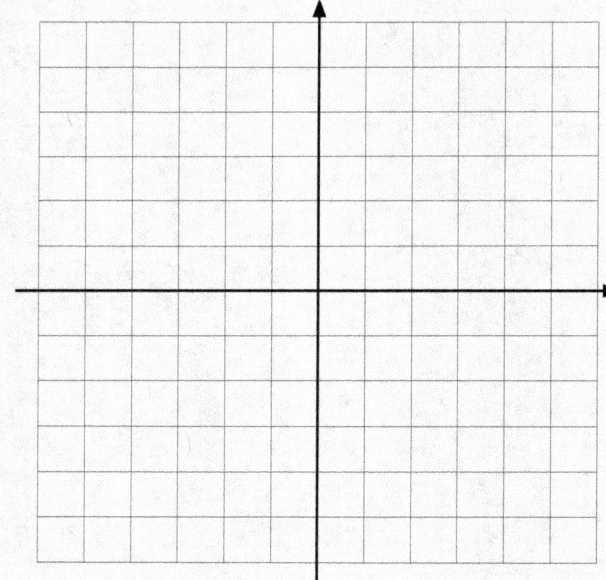

2.8 Linear Inequalities in Two Variables

Horizontal or Vertical Line Boundary

When working with inequalities that have vertical or horizontal lines as boundaries, we should first isolate the variable, x or y, to the left side of the inequality.

- Graph the boundary line by replacing the linear inequality with a linear equality.

    ◇ Use a **dashed line** if the inequality is strictly less than or strictly greater than (that is, < or >). This means that points on the line are not included in the solution set.

    ◇ Use a **solid line** if the inequality is not strict (that is, $\leq$ or $\geq$). This means that points on the line are included in the solution set.

- Shading when the variable involved is y

    ◇ Shade **above** the horizontal line if the inequality is greater than (>) or greater than or equal to ($\geq$).

    ◇ Shade **below** the horizontal line if the inequality is less than (<) or less than or equal to ($\leq$).

- Shading when the variable involved is x

    ◇ Shade to the **right** of the vertical line if the inequality is greater than (>) or greater than or equal to ($\geq$).

    ◇ Shade to the **left** of the vertical line if the inequality is less than (<) or less than or equal to ($\leq$).

Example 3 Graph the set of solutions to the following linear inequality.

a) $y \geq -2$

b) $x - 4 < 0$

Solution.

a) The variable y is already isolated to the left.

Graph $y = -2$. This is a horizontal line and it is the boundary line.

Since the inequality is non-strict ($\geqslant$), the boundary line will be a solid line and we will shade above the line.

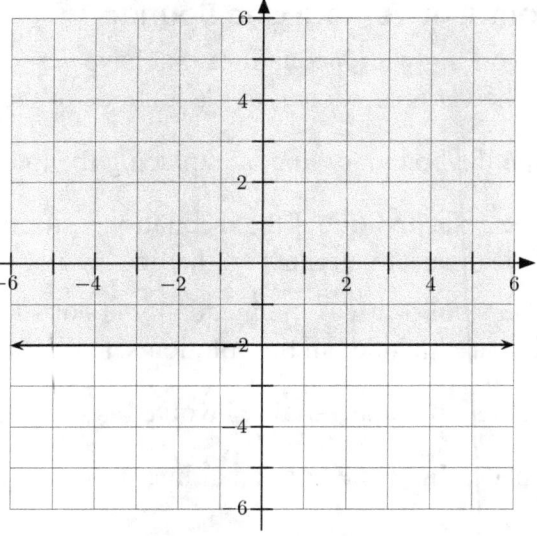

b) We will need to first isolate x.

$$\begin{aligned} x - 4 &< 0 \\ +4 &+4 \\ \hline x &< 4 \end{aligned}$$

Add 4 to each side

Inequality with x on the left

Graph $x = 4$. This is a vertical line and it is the boundary line.

Since the inequality is strictly less than ($<$), the boundary line is a dashed line and we will shade to the right of the line.

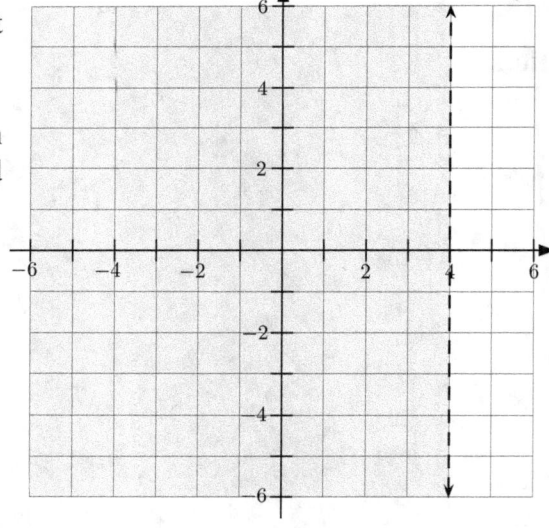

2.8 Linear Inequalities in Two Variables

Exercise 5 Class Example
Graph the set of solutions to the following linear inequality.

a) $y - 3 > 2$

b) $x + 1 \leqslant 0$

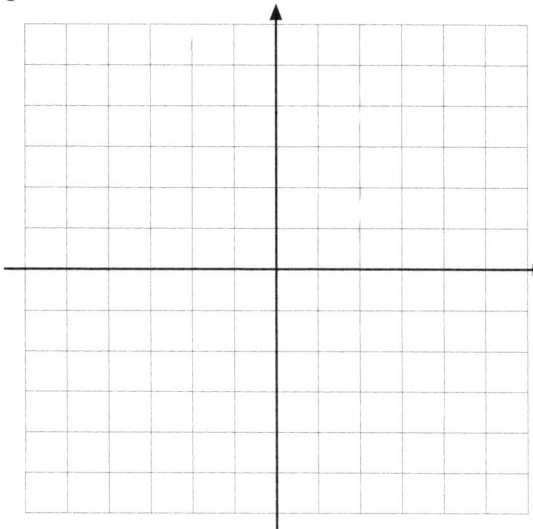
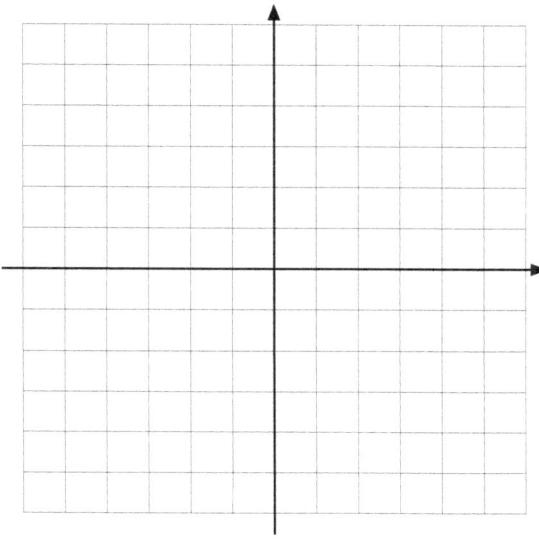

Exercise 6 You Try
Graph the set of solutions to the following linear inequality.

a) $3 < x$

b) $5 + y \leqslant 7$

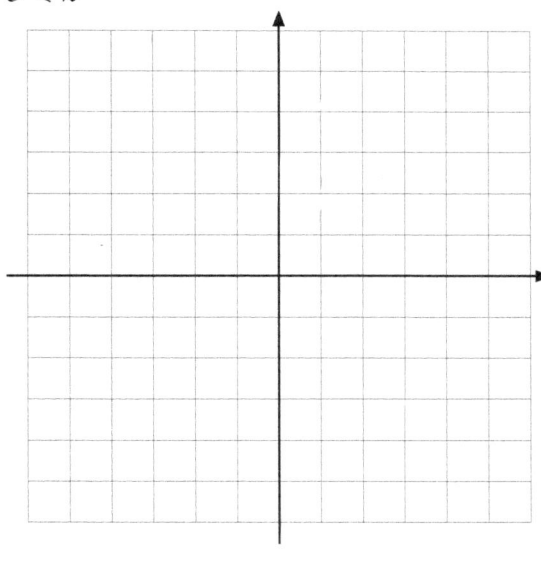
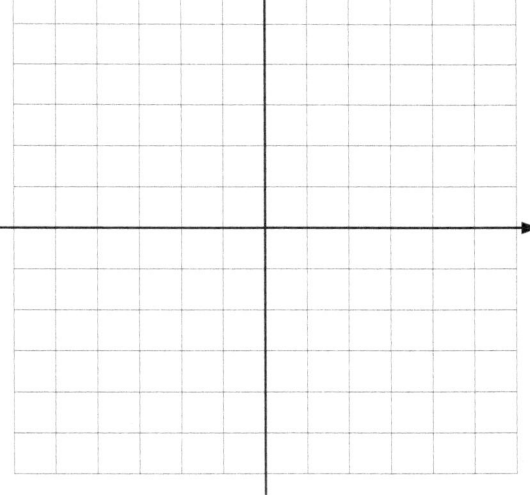

2.8: Exercises

Graph the set of solutions to the inequality.

1. $y \geq -\frac{1}{3}x - 2$

2. $y \leq \frac{2}{3}x$

3. $y < 3x - 1$

4. $x < y$

5. $y \geq -2$

6. $3 < x$

7. $y + 4x \leq 5$

8. $x - y \leq -1$

9. $3x + y \geq 0$

10. $2y < \frac{1}{2}x$

11. $-x + 3y \geq 6$

12. $x - 2y > -4$

13. $y - 4 < 0$

14. $x + 1 < -1$

Chapter 2 Assessment

1. Graph the following lines by first producing a table of at least 3 solutions.

   (a) $2x - y = 5$
   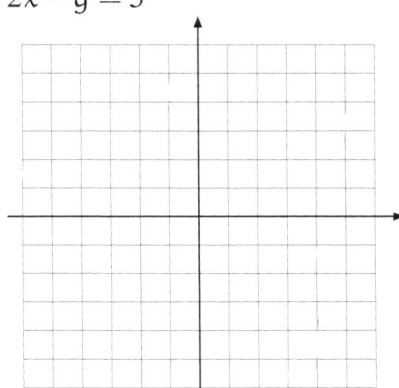

   (b) $x = -2$
   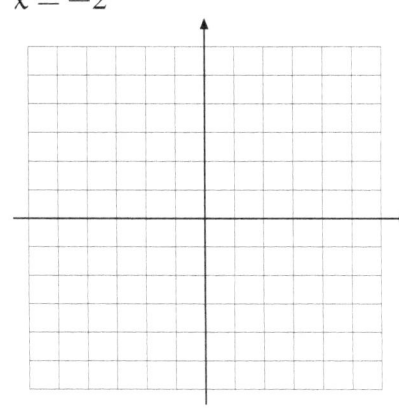

2. For each equation, find the x-intercept and y-intercept. Write your answer as an ordered pair. Then graph the line using the intercepts found.

   a) $y = \frac{1}{2}x - 1$
   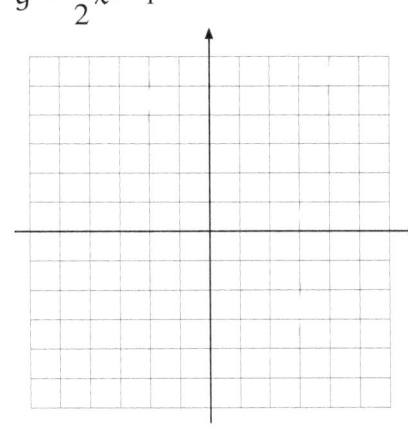

   b) $2x + y = -6$
   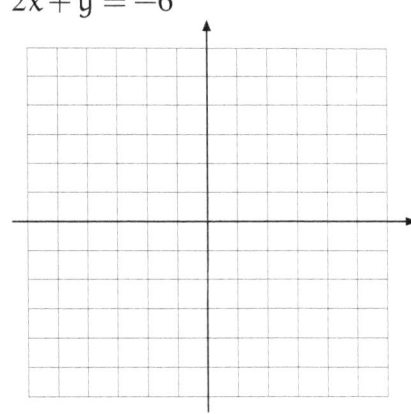

3. For each equation, identify the slope and y-intercept. Graph the line using the information found.

   a) $3x + y = 0$
   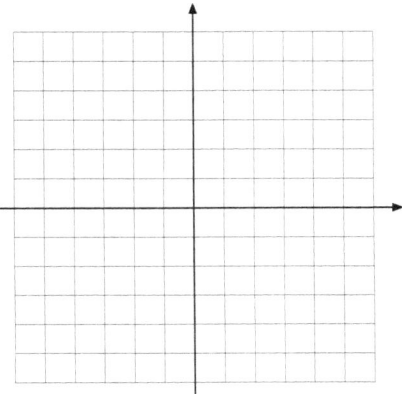

b) $y = 3$

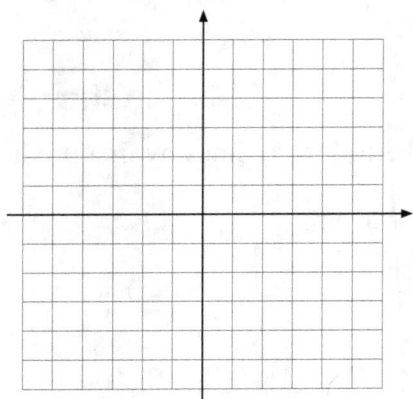

4. Sketch the graph of each inequality.

   a) $y > \dfrac{2}{3}x - 4$

   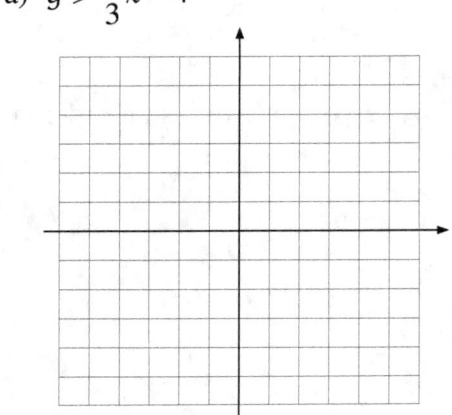

   b) $x \geq y$

   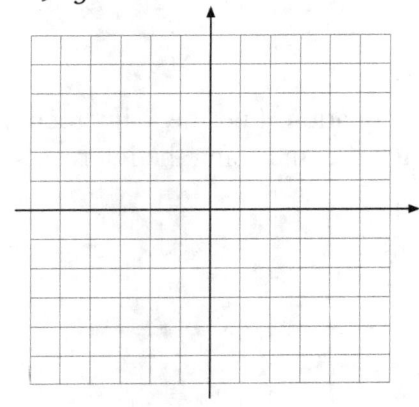

5. Find the slope of the line though each pair of points.

   a) $(0, -3)$ and $(4, 3)$

   b) $(-3, -7)$ and $(9, -5)$

6. Find the equation of a line with the following properties. Write the answer in slope-intercept form.

   a) A line has slope $-\dfrac{5}{2}$ and y-intercept $(0, 6)$

   b) A line goes through $(6, -3)$ and $(4, -2)$

2.8 Linear Inequalities in Two Variables

c) A horizontal line going through $(5,-8)$

d) A line goes through $(9,-3)$ and is parallel to $2x+y=1$

e) A line goes through $(1,-7)$ and is perpendicular to $x-3y=4$

3. Systems of Linear Equations

3.1 Introduction to Systems of Linear Equations

Objective: To determine if a given point is a solution of a system of equations

Example 1 A car-stereo warehouse has two options of compensation for their employees.

- Plan A pays $200 a week plus a $30 commission per stereo sold.

- Plan B has no salary but pays a $50 commission per stereo sold.

Paul is employed under Plan A and Bob is employed under Plan B. Last week, Paul and Bob were both paid the same amount (before deductions). How many stereos did each one of them sell last week?

Solution.
Notice that each plan can be represented by a linear equation as follows.

$$y = 30x + 200 \quad \text{Plan A}$$
$$y = 50x \quad \text{Plan B}$$

where x is the number of stereos sold last week, and y is the total salary paid last week.

The problem is asking to find the value of x that will produce the same y-value for both equations. Let's explore some values for each equation.

Plan A

x	y
1	230
2	260
3	290
5	350
10	500

Plan B

x	y
1	50
2	100
3	150
5	250
10	500

We found our solution. Each one sold 10 car stereos last week. Notice that the point (10, 500) is a solution for **both equations**.

Plan A

$$500 \stackrel{?}{=} 30(10) + 200$$
$$500 = 300 + 200 \checkmark$$

Plan B

$$y \stackrel{?}{=} 50(10)$$
$$y = 500 \checkmark$$

Systems of Two Linear Equations

In situations as above, we will call the set of two equations

$$\begin{cases} y = 30x + 200 \\ y = 50x \end{cases}$$

a **system of two linear equations**. The point (10, 500) is a **solution** to this system.

In this chapter, we will learn more about systems of two linear equations, their solutions and how to find them. The method presented above is not the best way to find a solution and will not work for two general linear equations.

Notice that since each equation represents a line in the xy-plane, a solution to a system of two linear equations, if it exists, will be exactly one point. It could also happen that the two lines do not have a common point, in which case we will say that the system has **no solution**.

Example 2 Determine if the point $(-2, 6)$ is a solution to the system of linear equations

$$\begin{cases} 3x + 2y = 6 \\ x - y = -8 \end{cases}$$

Solution.
In order to be a solution to the system, the point $(-2, 6)$ must be a solution to each line. Let's check. Substitute the ordered pair $(x, y) = (-2, 6)$ into each linear equation.

3.1 Introduction to Systems of Linear Equations

First Equation:

$$3x + 2y = 6$$
$$3(-2) + 2(6) \stackrel{?}{=} 6$$
$$-6 + 12 = 6 \checkmark$$

Second Equation:

$$x - y = -8$$
$$-2 - 6 \stackrel{?}{=} -8$$
$$-8 = -8 \checkmark$$

So indeed, the point $(-2, 6)$ is a solution to the system.

Example 3 Determine if the point $(-2, -4)$ is a solution to the system of linear equations

$$\begin{cases} 5x - 4y = 6 \\ 5x - 2y = 2 \end{cases}$$

Solution.
In order to be a solution to the system, the point $(-2, -4)$ must be a solution to each line. Let's check. Substitute $x = -2$ and $y = -4$ into each linear equation.

First Equation:

$$5x - 4y = 6$$
$$5(-2) - 4(-4) \stackrel{?}{=} 6$$
$$-10 + 16 = 6 \checkmark$$

Second Equation:

$$5x - 2y = 2$$
$$5(-2) - 2(-4) \stackrel{?}{=} 2$$
$$-10 + 8 \neq 2$$

So the point $(-2, -4)$ is not a solution to the system.

Note. In the last example, we do not claim that the system does not have a solution. In fact, it does. We are only saying that the point $(-2, -4)$ is not the solution. In the following sections, we will learn methods to find the solution for this kind of problems.

Exercise 1 Class Example

Determine if the point $(-5, 13)$ is a solution to the system of equations $\begin{cases} 4x + 2y = 6 \\ x + y = 8 \end{cases}$

Exercise 2 You Try

Determine if the point $(5,7)$ is a solution to the system of equations $\begin{cases} -x+y=2 \\ -6x+10y=40 \end{cases}$

Example 4 Determine if the point $\left(\dfrac{31}{10}, \dfrac{3}{5}\right)$ is a solution to the system of linear equations

$$\begin{cases} 2x-2y=5 \\ 2x+3y=8 \end{cases}$$

Solution.

In order to be a solution to the system, the point $\left(\dfrac{31}{10}, \dfrac{3}{5}\right)$ must be a solution to each line. Let's check. Substitute $x=\dfrac{31}{10}$ and $y=\dfrac{3}{5}$ into each linear equation.

First equation:

$$2x-2y=5$$
$$2\left(\dfrac{31}{10}\right)-2\left(\dfrac{3}{5}\right)\stackrel{?}{=}5$$
$$\dfrac{31}{5}-\dfrac{6}{5}\stackrel{?}{=}5$$
$$\dfrac{25}{5}=5 \checkmark$$

Second equation:

$$2x+3y=8$$
$$2\left(\dfrac{31}{10}\right)+3\left(\dfrac{3}{5}\right)\stackrel{?}{=}8$$
$$\dfrac{31}{5}+\dfrac{9}{5}\stackrel{?}{=}8$$
$$\dfrac{40}{5}=8 \checkmark$$

So the point $\left(\dfrac{31}{10}, \dfrac{3}{5}\right)$ is a solution to the system.

3.1 Introduction to Systems of Linear Equations

Exercise 3 Class Example

Determine if the point $\left(-\dfrac{7}{3}, \dfrac{5}{3}\right)$ is a solution to the system of equations $\begin{cases} 4x+2y=6 \\ x+y=8 \end{cases}$

Exercise 4 You Try

Determine if the point $\left(\dfrac{1}{4}, -\dfrac{9}{4}\right)$ is a solution to the system of equations $\begin{cases} 5x-3y=8 \\ -x-y=2 \end{cases}$

3.1: Exercises

Determine if the given point is a solution to the corresponding system of equations.

1. $(2,1)$;
$3x - y = 5$
$2x + 3y = 7$

2. $(2,-8)$;
$y = -4x$
$x - y = 10$

3. $(2,-2)$;
$3x - 2y = 10$
$4x + 5y = 8$

4. $(6,-2)$;
$x + y = 4$
$x - 3y = 12$

5. $\left(\dfrac{7}{2}, \dfrac{11}{2}\right)$;
$x - y = -2$
$3x - y = 9$

6. $\left(-19, -\dfrac{85}{2}\right)$;
$x - 2y = 66$
$3x - 2y = 28$

7. $(2,-2)$;
$3x - 2y = 10$
$4x + 5y = 8$

8. $(10,10)$;
$x = 2y - 10$
$y = -15 + 3x$

9. $\left(-\dfrac{1}{2}, -\dfrac{5}{4}\right)$;
$-x - 2y = 3$
$x - 2y = 2$

10. $(4,-5)$;
$4x - 5y = 41$
$-3x + 2y = 18$

3.2 Solving Systems of Linear Equations by Graphing

Objective: To solve systems of linear equations by graphing

In this section, we will find the solutions to systems of two linear equations by graphing both lines and finding their common point of intersection. This method does not always work but it is a good starting point to understand systems of linear equations.

Example 1

Find the point where the two lines intersect by graphing both lines: $\begin{cases} 3x - y = 5 \\ x + y = 3 \end{cases}$

Solution.
Let us first rewrite the equation of the lines in slope-intercept form.

First line:

$$3x - y = 5$$
$$\underline{-3x \qquad -3x} \qquad \text{Subtract 3x from each side}$$
$$-y = -3x + 5$$
$$\frac{-y}{-1} = \frac{-3x}{-1} + \frac{5}{-1} \qquad \text{Divide each term by } -1$$
$$y = 3x - 5 \qquad \text{First equation in slope-intercept form}$$

The first line has slope, $m = 3$ and y-intercept, $(0, -5)$.

Second line:

$$x + y = 3$$
$$\underline{-x \qquad -x} \qquad \text{Subtract x from each side}$$
$$y = -x + 3 \qquad \text{Second equation in slope-intercept form}$$

The second line has slope, $m = -1$ and y-intercept, $(0, 3)$. We can now graph both lines on the same plane.

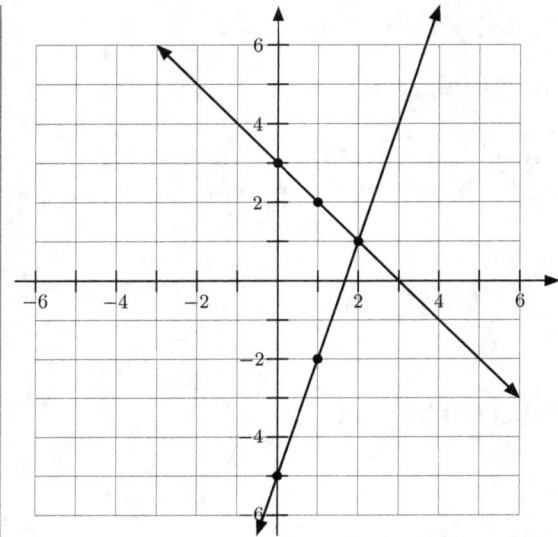

Since the point $(2,1)$ belongs to both lines, it is a solution to each equation. Verify that we have the correct solution by substituting $(x,y) = (2,1)$ into each of the original equations.

First Equation:	Second Equation:
$3x - y = 5$	$x + y = 3$
$3(2) - (1) \stackrel{?}{=} 5$	$(2) + (1) \stackrel{?}{=} 3$
$6 - 1 = 5$ ✓	$3 = 3$ ✓

The point $(2, 1)$ is a solution to the system of linear equations. It is also the point where the two lines intersect.

In the above example, the *system of linear equations* is the set of two original equations, $3x - y = 5$ and $x + y = 3$. The *solution* to the system of linear equations is the pair $(x, y) = (2, 1)$.

3.2 Solving Systems of Linear Equations by Graphing

Exercise 1 **Class Example**

Find the point where the two lines intersect by graphing both lines: $\begin{cases} -2x + y = -3 \\ x - y = 3 \end{cases}$

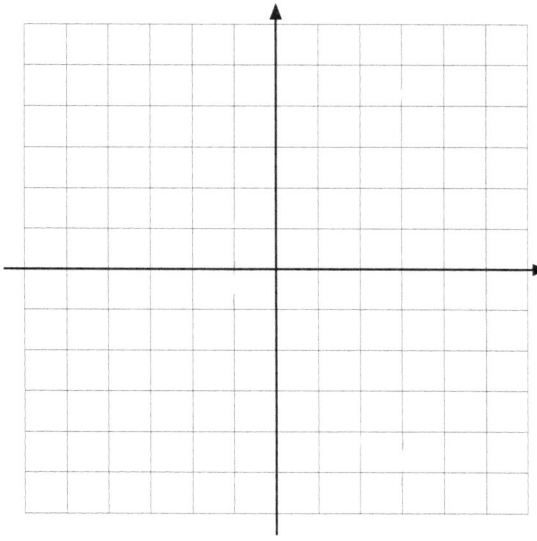

Exercise 2 **You Try**

Find the point where the two lines intersect by graphing both lines: $\begin{cases} x + y = 0 \\ 2x + y = 3 \end{cases}$

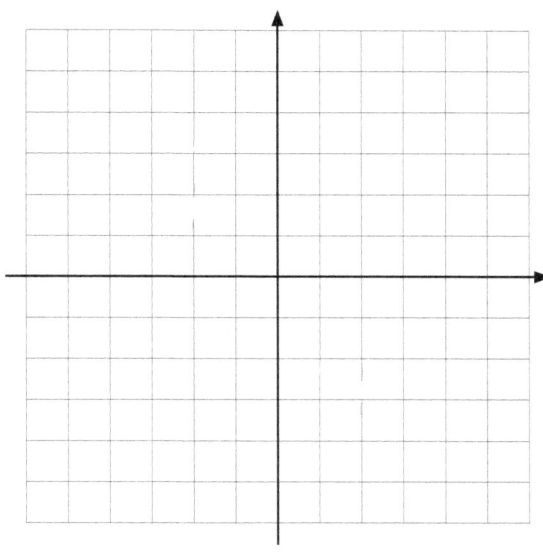

Exercise 3 You Try

Find the point where the two lines intersect by graphing both lines: $\begin{cases} x + 3y = 6 \\ x - 2y = -4 \end{cases}$

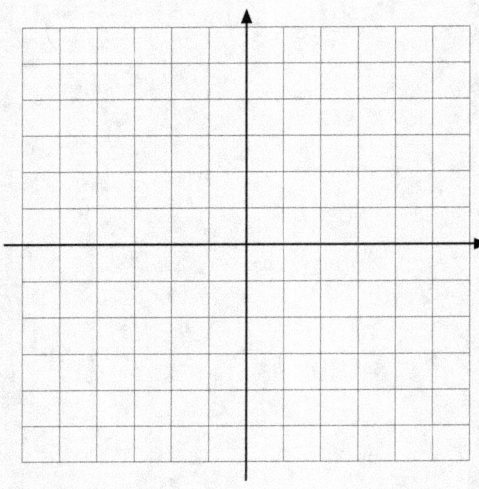

Example 2

Solve the system of linear equations by graphing: $\begin{cases} 2x - 3y = 12 \\ 4x - 6y = -6 \end{cases}$

Solution.
Let us first rewrite the equation of the lines in slope-intercept form.

First line:

$$2x - 3y = 12$$
$$\underline{-2x \qquad -2x} \qquad \text{Subtract 2x from each side}$$
$$-3y = -2x + 12$$

$$\frac{-3y}{-3} = \frac{-2x}{-3} + \frac{12}{-3} \qquad \text{Divide each term by } -3$$

$$y = \frac{2}{3}x - 4 \qquad \text{First equation in slope-intercept form}$$

The first line has slope, $m = \dfrac{2}{3}$ and y-intercept, $(0, -4)$.

3.2 Solving Systems of Linear Equations by Graphing

Second line:

$$4x - 6y = -6$$
$$\underline{-4x \qquad\quad -4x} \qquad \text{Subtract 4x from each side}$$
$$-6y = -4x - 6$$

$$\frac{-6y}{-6} = \frac{-4x}{-6} - \frac{6}{-6} \qquad \text{Divide each term by } -6$$

$$y = \frac{2}{3}x + 1 \qquad \text{Second equation in slope-intercept form}$$

The second line has slope, $m = \dfrac{2}{3}$ and y-intercept, $(0, 1)$.

Notice that both lines have the same slope, $m = \dfrac{2}{3}$ but different y-intercepts. Let us take a look at the graph of both lines.

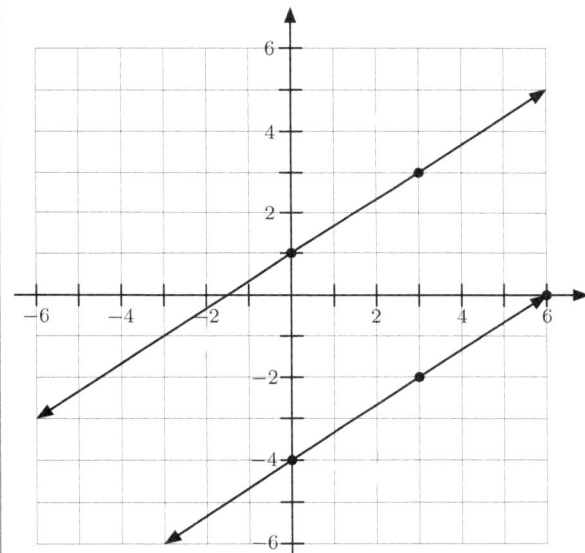

We see that the two lines do not intersect.

In this case, we say that there is **no solution** to the system of linear equations and can be represented by the symbol ∅. The symbol, ∅, is used in mathematics to denote the *empty set*, the set that has no elements. Since the solution to the above system of linear equations has *no elements*, we can refer to the solution as being empty, or ∅.

Exercise 4 Class Example

Solve the system of linear equations by graphing: $\begin{cases} 2x + 5y = -10 \\ 8x + 20y = -40 \end{cases}$

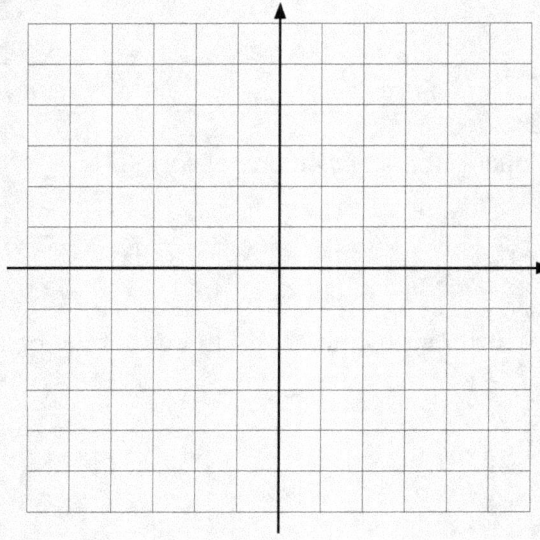

Exercise 5 You Try

Solve the system of linear equations by graphing: $\begin{cases} x + 3y = 15 \\ 2x + 6y = 24 \end{cases}$

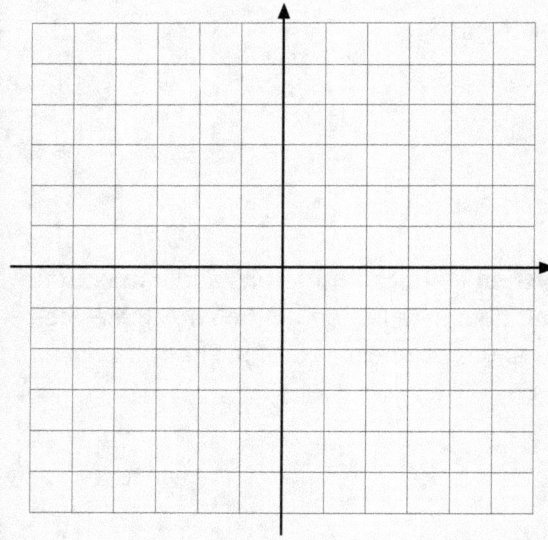

3.2 Solving Systems of Linear Equations by Graphing

Example 3

Solve the system of linear equations by graphing: $\begin{cases} 4x + 3y = -12 \\ -8x - 6y = 24 \end{cases}$

Solution.

Let us first rewrite the equation of the lines in slope-intercept form.

First line:

$$4x + 3y = -12$$
$$\underline{-4x \qquad\qquad -4x} \qquad \text{Subtract 4x from each side}$$
$$3y = -4x - 12$$

$$\frac{3y}{3} = \frac{-4x}{3} - \frac{12}{3} \qquad \text{Divide each term by 3}$$

$$y = -\frac{4}{3}x - 4 \qquad \text{First equation in slope-intercept form}$$

The first line has slope, $m = -\dfrac{4}{3}$ and y-intercept, $(0, -4)$.

Second line:

$$-8x - 6y = 24$$
$$\underline{+8x \qquad\qquad +8x} \qquad \text{Add 8x to each side}$$
$$-6y = 8x + 24$$

$$\frac{-6y}{-6} = \frac{8x}{-6} + \frac{24}{-6} \qquad \text{Divide each term by } -6$$

$$y = -\frac{4}{3}x - 4 \qquad \text{Second equation in slope-intercept form}$$

The second line also has slope, $m = -\dfrac{4}{3}$ and y-intercept, $(0, -4)$. Both equations represent the same line!

Notice that any point on the first line is also on the second line and vice versa. That is why we see only one line graphed.

In this case, we say that there are **infinitely many solutions** to the system of linear equations.

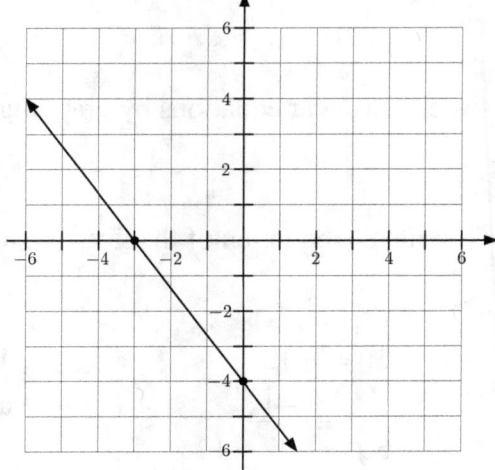

3.2 Solving Systems of Linear Equations by Graphing

Exercise 6 **Class Example**

Solve the system of linear equations by graphing: $\begin{cases} 2x + 3y = 9 \\ -6x = 9y - 27 \end{cases}$

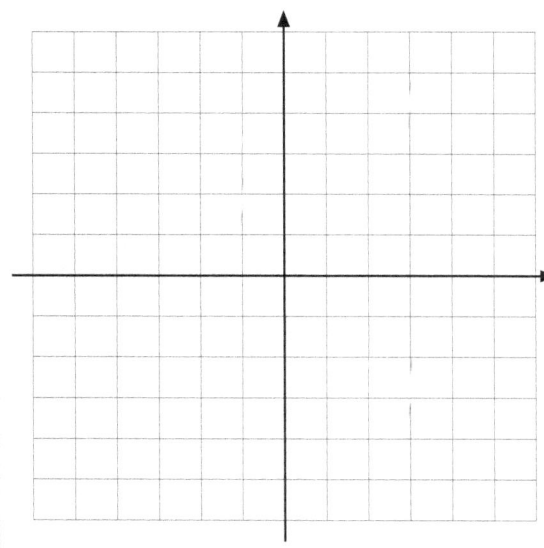

Exercise 7 **You Try**

Solve the system of linear equations by graphing: $\begin{cases} 2y = 3x + 2 \\ 6x = 4y - 4 \end{cases}$

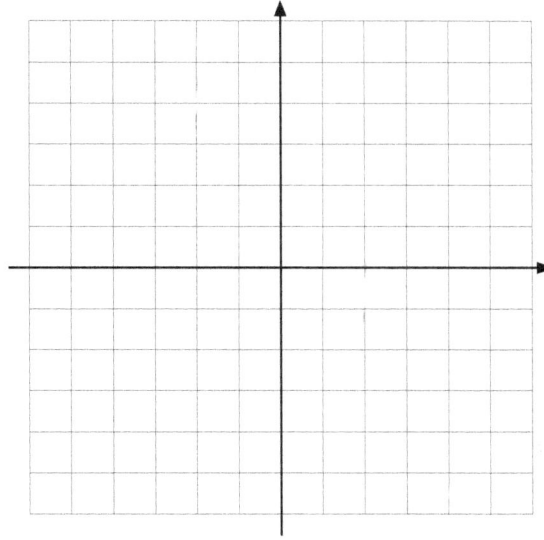

Sometimes, the graphing method is not the best way to solve a system of linear equations because the two lines can intersect at a point that will not be easy to find from a graph. Let us take a look at some of the difficulties with graphing.

Exercise 8 Class Example

Solve the system of linear equations by graphing: $\begin{cases} x+y=2 \\ x-2y=1 \end{cases}$

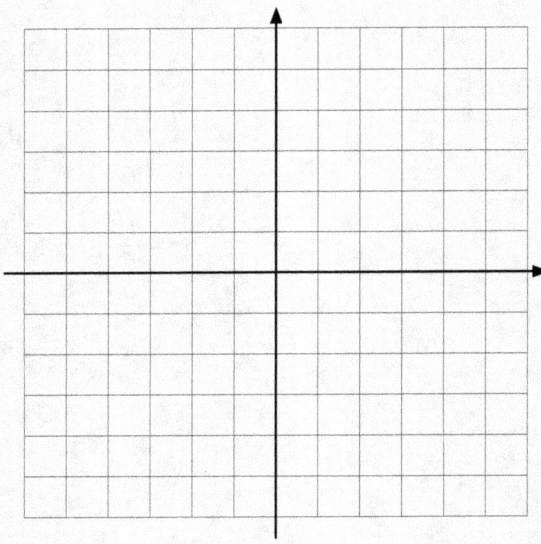

Exercise 9 You Try

Solve the system of linear equations by graphing: $\begin{cases} x-y=7 \\ 2y=x+4 \end{cases}$

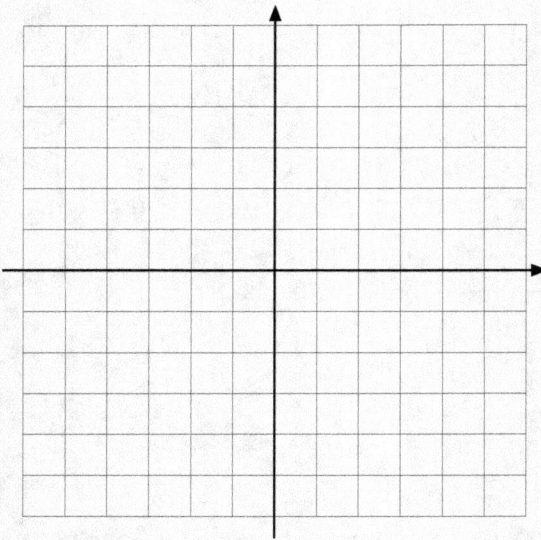

3.2 Solving Systems of Linear Equations by Graphing

In the next sections, we will learn other methods to solve a system that do not depend on the graphs of the lines.

> **World View Note** Persian mathematician, Omar Khayyam would solve algebraic problems geometrically by intersecting graphs rather than solving them algebraically.

3.2: Exercises

Solve the following systems of linear equations by graphing.

1. $\begin{array}{l} -x+y=1 \\ -3x+y=-1 \end{array}$

2. $\begin{array}{l} -x+2y=8 \\ 2x+y=-1 \end{array}$

3. $\begin{array}{l} y=\dfrac{x}{2}+1 \\ x-y=-3 \end{array}$

4. $\begin{array}{l} -x+y=1 \\ 2x-2y=3 \end{array}$

5. $\begin{array}{l} x+3y=0 \\ x-y=4 \end{array}$

6. $\begin{array}{l} 5x+3y=6 \\ 2-y=\dfrac{5}{3}x \end{array}$

7. $\begin{array}{l} 3x-y=4 \\ x-2y=-2 \end{array}$

8. $\begin{array}{l} y=\dfrac{2}{3}x-4 \\ x+y=1 \end{array}$

9. $\begin{array}{l} 4y=3x-12 \\ 2x+8y=8 \end{array}$

10. $\begin{array}{l} x-y=3 \\ y=-2 \end{array}$

3.3 Substitution Method

Objective: To solve systems of linear equations by substitution

Solving systems of linear equations by graphing has limitations. First, it requires the graph to be perfectly drawn. If the lines are not perfectly drawn, we may arrive at the wrong answer. Second, graphing is not a great method to use if the answer contains a decimal or fraction. For these reasons, we need a different approach to solving systems of linear equations.

In this section, we will learn the algebraic approach called **substitution.** The strategy is as follows.

1. Solve one of the equations for one of the variables (we call this equation the "substitution equation")

2. Substitute the expression obtained if Step 1 into the *other equation.* This should produce an equation in one variable.

3. Solve the equation obtained in Step 2.

4. Substitute the value obtained in Step 3, into the "substitution equation" and solve for the other variable.

5. Check your solution in the original equations.

Example 1 Given the equation $y = 2x - 3$, solve for y when $x = 5$.

Solution.

$y = 2x - 3$ Substitute x with 5
$y = 2(5) - 3$ Multiply first
$y = 10 - 3$ Subtract
$y = 7$ Our Solution

When one variable is expressed in terms of another variable, we can substitute that expression in for the variable in the other equation. It is very important to use parenthesis when we substitute.

One Isolated Variable

Example 2

Solve the following system of linear equations and check your answer: $\begin{cases} 2x - 3y = 7 \\ y = 3x - 7 \end{cases}$

Solution.
In the second equation, y is expressed in terms of x. We can substitute $3x - 7$ for y in the first

equation.

$$2x - 3(3x - 7) = 7 \quad \text{Distribute } -3$$
$$2x - 9x + 21 = 7 \quad \text{Combine like terms}$$
$$-7x + 21 = 7 \quad \text{Subtract 21 from each side}$$
$$-7x = -14 \quad \text{Divide by } -7$$
$$x = 2 \quad \text{We have our x!}$$

$$y = 3x - 7$$
$$y = 3(2) - 7 \quad \text{Substitute } x = 2 \text{ into the equation to solve for y}$$
$$y = 6 - 7 \quad \text{Simplify}$$
$$y = -1 \quad \text{We have our y!}$$
$$(2, -1) \quad \text{Our Solution}$$

Verify that we have the correct solution by substituting $(x, y) = (2, -1)$ into each of the original equations.

First Equation:
$$2x - 3y = 7$$
$$2(2) - 3(-1) \stackrel{?}{=} 7$$
$$4 + 3 = 7 \checkmark$$

Second Equation:
$$y = 3x - 7$$
$$-1 \stackrel{?}{=} 3(2) - 7$$
$$-1 = 6 - 7 \checkmark$$

So the point $(2, -1)$ is a solution to the system of linear equation.

Exercise 1 Class Example

Solve the following system of linear equations and check your answer: $\begin{cases} y = x \\ x + y = 6 \end{cases}$

3.3 Substitution Method

Exercise 2 You Try
Solve the following system of linear equations. Be sure to check your answer.

a) $\begin{cases} y = 2x + 5 \\ y - 4x = 4 \end{cases}$

b) $\begin{cases} y = x + 3 \\ x - \dfrac{1}{2}y = -2 \end{cases}$

c) $\begin{cases} 3x - 4y = 9 \\ x = y + 3 \end{cases}$

No Isolated Variable

Sometimes there is no isolated variable in either equation, so we need to first isolate a variable. Look for an equation where one of the variable has a coefficient of 1. Use that equation and express the variable with a coefficient of 1 in terms of the other variable.

Example 3

Solve the following system of linear equations and check your answer: $\begin{cases} 3x + 2y = 1 \\ x - 5y = 6 \end{cases}$

Solution.
In the first equation, neither x nor y has a coefficient of 1. However, in the second equation, we see that x has a coefficient of 1. Solve for x in terms of y.

$$x - 5y = 6 \quad \text{Second equation}$$
$$x - 5y + 5y = 6 + 5y \quad \text{Add 5y to each side}$$
$$x = 6 + 5y$$

Substitute $x = 6 + 5y$ into the other equation, $3x + 2y = 1$.

$$3(6 + 5y) + 2y = 1 \quad \text{Distribute 3}$$
$$18 + 15y + 2y = 1 \quad \text{Combine like terms}$$
$$18 + 17y = 1 \quad \text{Subtract 18 from each side}$$
$$17y = -17 \quad \text{Divide by 17}$$
$$y = -1 \quad \text{We have our y!}$$

$$x = 6 + 5(-1) \quad \text{Substitute } y = -1 \text{ into } x = 6 + 5y \text{ to solve for y}$$
$$x = 6 - 5 \quad \text{Simplify}$$
$$x = 1 \quad \text{We have our x!}$$
$$(1, -1) \quad \text{Our Solution}$$

Verify that we have the correct solution by substituting $(x, y) = (1, -1)$ into each of the original equations.

First Equation:
$$3x + 2y = 1$$
$$3(1) + 2(-1) \stackrel{?}{=} 1$$
$$3 - 2 = 1 \checkmark$$

Second Equation:
$$x - 5y = 6$$
$$(1) - 5(-1) \stackrel{?}{=} 6$$
$$1 + 5 = 6 \checkmark$$

So the point $(1, -1)$ is a solution to the system of linear equations.

3.3 Substitution Method

Exercise 3 Class Example

Solve the following system of linear equations and check your solution: $\begin{cases} 3x - 2y = 2 \\ x + 4y = 3 \end{cases}$

Exercise 4 You Try
Solve the following system of linear equations. Be sure to check your answer.

a) $\begin{cases} x + y = 5 \\ x - y = -1 \end{cases}$

b) $\begin{cases} \dfrac{1}{2}x + 3y = -2 \\ x + 2y = 4 \end{cases}$

Sometimes, none of the equations have a lone variable and neither of the variables have a coefficient of one. In order to solve the system of linear equations by substitution, you will need to express one variable in terms of another. Choose your variable wisely.

Example 4

Solve the following system of linear equations: $\begin{cases} 5x - 6y = -14 \\ -2x + 4y = 12 \end{cases}$

Be sure to check your answer.
Solution.
None of the equations has a lone variable. We will solve for x in the second equation because the coefficients of x and y, including the constant terms are all divisible by -2, the coefficient of x in the second equation.

$-2x + 4y = 12$	Divide each term by -2
$x - 2y = -6$	Add $2y$ to each side of the equation
$x = 2y - 6$	

Substitute $x = 2y - 6$ into the first equation and solve for y.

$5x - 6y = -14$	
$5(2y - 6) - 6y = -14$	Substitute $x = 2y - 6$ and distribute
$10y - 30 - 6y = -14$	Combine like terms
$4y - 30 = -14$	Add 30 to each side
$4y = 16$	Divide each side by 4
$y = 4$	We have our y!
$x = 2(4) - 6$	Substitute $y = 4$ into the the equation $x = 2y - 6$
$x = 8 - 6$	Simplify
$x = 2$	We have our x!
$(2, 4)$	Our Solution

Verify that we have the correct solution by substituting $(x, y) = (2, 4)$ into each of the original equations.

First Equation:
$$5x - 6y = -14$$
$$5(2) - 6(4) \stackrel{?}{=} -14$$
$$10 - 24 = -14 \checkmark$$

Second Equation:
$$-2x + 4y = 12$$
$$-2(2) + 4(4) \stackrel{?}{=} 12$$
$$-4 + 16 = 12 \checkmark$$

So the point $(2, 4)$ is a solution to the system of linear equations.

Exercise 5 Class Example

Solve the following system of linear equations and check your answer: $\begin{cases} 2x - 4y = 8 \\ 9x + 3y = -6 \end{cases}$

Exercise 6 You Try

Solve the following system of linear equations and check your answer: $\begin{cases} 2y - 8x = 6 \\ 2x + 3y = 2 \end{cases}$

As we saw in graphing, it is possible that a system of linear equations has no solution, ∅. This occurs when two lines are parallel to each other. It is also possible for a system to have infinitely many solutions. This happens when we have the same line expressed as two equivalent equations. When solving these kinds of systems algebraically, the process takes an interesting turn.

Example 5

Solve the following system of linear equations and check your answer: $\begin{cases} 6x - 3y = -9 \\ y = 2x + 5 \end{cases}$

Solution.
Substitute the second equation, $y = 2x + 5$ into the first equation, $6x - 3y = -9$.

$6x - 3(2x + 5) = -9$	Substitute $y = 2x + 5$ and distribute
$6x - 6x - 15 = -9$	Combine like terms
$-15 \neq -9$	Variables are gone! False statement.
No Solution or ∅	Our Solution

Because we have a false statement, we know that nothing will work in both equations.

Example 6

Solve the following system of linear equations and check your answer: $\begin{cases} x + 2y = 4 \\ 3x = 12 - 6y \end{cases}$

Solution.
None of the equations has a lone variable. We will solve for x in the second equation.

$3x = 12 - 6y$	Divide each term by 3
$x = 4 - 2y$	We have our x

Substitute $x = 4 - 2y$ into the first equation, $x + 2y = 4$.

$x + 2y = 4$	First equation
$(4 - 2y) + 2y = 4$	Substitute $x = 4 - 2y$
$4 - 2y + 2y = 4$	Combine like terms
$4 = 4$	Variables are gone! True statement.
Infinitely many solutions	Our Solution

Because the variables are gone and we have a true statement, the two lines must be one and the same. Any point on the first line will also be on the second line. Since lines have an infinite number of points, we have infinitely many solutions

3.3 Substitution Method

Exercise 7 Class Example
Solve the following system of linear equations. Be sure to check your answer.

a) $\begin{cases} y = 2x - 5 \\ 4x - 2y = 10 \end{cases}$

b) $\begin{cases} \frac{1}{2}x - y = -2 \\ x = 2y \end{cases}$

Exercise 8 You Try
Solve the following system of linear equations. Be sure to check your answer.

a) $\begin{cases} y = 4 - 2x \\ 6x + 3y = 12 \end{cases}$

b) $\begin{cases} x = -y \\ y = 4 - x \end{cases}$

3.3: Exercises

Solve the following systems of linear equations by substitution. Be sure to check your answers.

1. $y = -3x$
   $y = 6x - 9$

2. $b = a + 5$
   $b = -2a - 4$

3. $q = -6p + 3$
   $q = 6p + 3$

4. $3h - g = 12$
   $g = 6h + 21$

5. $6m - 4n = -8$
   $n = -6m + 2$

6. $-2x + 2y = 18$
   $y = 7x + 15$

7. $m = n - 4$
   $3m - 4n = -19$

8. $d = -8c + 19$
   $c - d = -16$

9. $2p + q = 8$
   $-7p - 6q = -8$

10. $a - b = 3$
    $-3a + 3b = 6$

11. $a - 5b = 7$
    $2a + 7b = -20$

12. $-2c - d = -5$
    $c - d = -23$

13. $3x + 2y = 4$
    $-4x + 2y = -6$

14. $-6x + y = 20$
    $18x - 3y = -18$

15. $7x + 5y = -13$
    $\frac{1}{4}x - y = -4$

16. $2a + b = 5$
    $4a - 5b = -4$

17. $x - 2y = -2$
    $y = \frac{1}{2}x + 1$

18. $x + 5y = 15$
    $-3x + 2y = 6$

19. $2m + 3n = -10$
    $7m + n = 3$

20. $2x + 6y = 2$
    $4y - 2y = -3$

21. $-6m + 6n = -12$
    $8m - 3n = 16$

22. $2a + 3b = 16$
    $-7a - b = 20$

3.3 Substitution Method

For each of the graphs below, do the following:

(a) Find the solution from the given graphs.

(b) Find the equation of each line.

(c) Using substitution, solve the system found in (b) to verify the solution found in (a).

23.
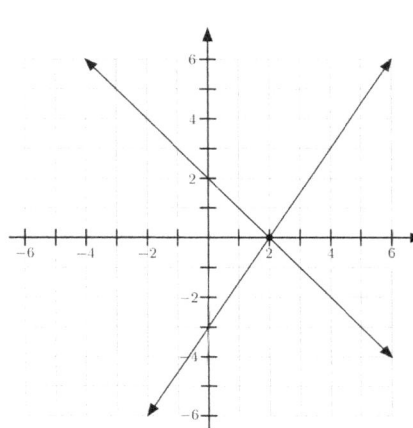

25.
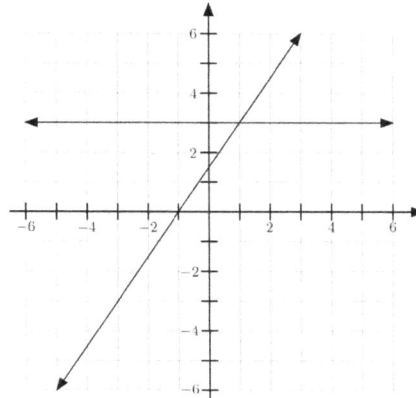

24.
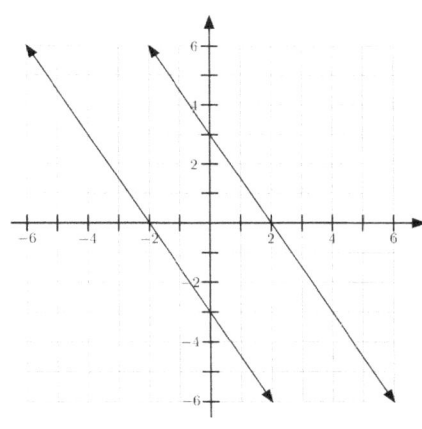

26.
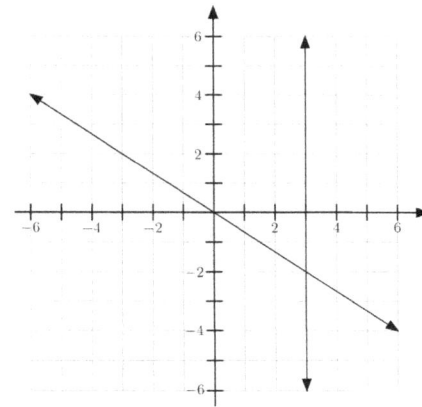

3.4 Elimination Method

Objective: To solve systems of linear equations using the elimination method

The third process used to solve a system of linear equations is the elimination method. This process is especially useful when the equations of the system are written in standard form $Ax+By=C$. The goal is to get the coefficients of one variable to be opposite of the other. The variable is eliminated by adding the two linear equations. This results in an equation in only one variable.

Example 1

Solve the following system of linear equations and check your answer: $\begin{cases} 3x-4y=8 \\ 5x+4y=-24 \end{cases}$

Solution.
Notice that the coefficients of the y-terms are opposite. We can eliminate the variable, y, by adding the two equations.

$$\begin{array}{r} 3x-4y=8 \\ 5x+4y=-24 \\ \hline 8x=-16 \end{array}$$

Solve the resulting linear equation in one variable.

$\quad 8x = -16 \quad\quad$ Divide each side by 8
$\quad x = -2 \quad\quad$ We have our x!

Substitute $x=-2$ into one of the original equations to solve for y. Here, we will substitute $x=-2$ into the second equation, $5x+4y=-24$.

$\quad 5(-2)+4y=-24 \quad\quad$ Simplify
$\quad -10+4y=-24 \quad\quad$ Add 10 to each side
$\quad 4y=-14 \quad\quad$ Divide each side by 4
$\quad y=-\dfrac{7}{2} \quad\quad$ We have our y!

$\quad \left(-2,-\dfrac{7}{2}\right) \quad\quad$ Our Solution

3.4 Elimination Method

Verify that we have the correct solution by substituting $(x,y) = \left(-2, -\frac{7}{2}\right)$ into each of the original equations.

First Equation:

$$3x - 4y = 8$$

$$3(-2) - 4\left(-\frac{7}{2}\right) \stackrel{?}{=} 8$$

$$-6 + 14 = 8 \checkmark$$

Second Equation:

$$5x + 4y = -24$$

$$5(-2) + 4\left(-\frac{7}{2}\right) \stackrel{?}{=} -24$$

$$-10 - 14 = -24 \checkmark$$

So the point $(2,-1)$ is a solution to the system of linear equations.

Exercise 1 Class Example

Solve the following system of linear equations and check your answer: $\begin{cases} 7a - b = 10 \\ 9a + b = 22 \end{cases}$

Exercise 2 You Try

Solve the following system of linear equations and check your answer: $\begin{cases} m - n = 4 \\ 2m + n = 8 \end{cases}$

In the previous examples, one variable was eliminated readily because the coefficients in the corresponding equations were opposites. We now examine a system in which the coefficients are not opposites. In this case, we re-balance our equations so that one of the two variables will eliminate. We can then follow the process demonstrated in the previous examples.

Example 2

Solve the following system of linear equations: $\begin{cases} -6x + 5y = 22 \\ 2x + 3y = 2 \end{cases}$. Be sure to check your answer.

Solution.
Notice that none of the variables have opposite coefficients. Choose a variable to eliminate. Here, we choose to eliminate x. Remember that you want the coefficients of the variable to be opposites. Multiply the second equation by 3. That way, one coefficient of x is −6 and the other is 6, so they are opposites. Note that when multiplying the second equation by 3, you must distribute the 3 to the entire equation.

3.4 Elimination Method

$$-6x + 5y = 22 \quad \Rightarrow \quad -6x + 5y = 22$$
$$3(2x + 3y) = 3(2) \quad \Rightarrow \quad \underline{6x + 9y = 6}$$
$$14y = 28$$

Solve the resulting linear equation in one variable.

$$14y = 28 \qquad \text{Divide each side by 14}$$
$$y = 2 \qquad \text{We have our y!}$$

Substitute $y = 2$ into one of the original equations to solve for x. Here, we will substitute $y = 2$ into the second equation, $2x + 3y = 2$.

$$2x + 3(2) = 2 \qquad \text{Simplify}$$
$$2x + 6 = 2 \qquad \text{Subtract 6 from each side}$$
$$2x = -4 \qquad \text{Divide each side by 2}$$
$$x = -2 \qquad \text{We have our x!}$$
$$(-2, 2) \qquad \text{Our Solution}$$

Verify that we have the correct solution by substituting $(x, y) = (-2, 2)$ into each of the original equations.

First Equation:

$$-6x + 5y = 22$$
$$-6(-2) + 5(2) \stackrel{?}{=} 22$$
$$12 + 10 = 22 \checkmark$$

Second Equation:

$$2x + 3y = 2$$
$$2(-2) + 3(2) \stackrel{?}{=} 2$$
$$-4 + 6 = 2 \checkmark$$

So the point $(-2, 2)$ is a solution to the system of linear equations.

Exercise 3 Class Example

Solve the following system of linear equations and check your answer: $\begin{cases} -6m - 2n = -9 \\ -\dfrac{3}{2}m + n = 3 \end{cases}$

Exercise 4 You Try

Solve the following system of linear equations and check your answer: $\begin{cases} \dfrac{4}{3}x - y = -2 \\ -4x + 5y = 20 \end{cases}$

3.4 Elimination Method

When we consider what number to use so that the equations have one set of variables that are opposite, we generally choose the Least Common Multiple (LCM) of the coefficients of the variable we want to eliminate.

Example 3

Solve the following system of linear equations and check your answer: $\begin{cases} -4x+5y=12 \\ -5x+3y=15 \end{cases}$

Solution.
First, choose which variable to eliminate. We choose to eliminate y. The LCM of the coefficients of y, 5 and 3, is 15. Multiply the first equation by 3 and the second equation by -5. That way, one coefficient of y is 15 and the other is -15, so they are opposites.

$$\begin{aligned} 3(-4x+5y) &= 3(12) &\Rightarrow& &-12x+15y &= 36 \\ -5(-5x+3y) &= -5(15) &\Rightarrow& &\underline{25x-15y} &= \underline{-75} \\ & & & & 13x &= -39 \end{aligned}$$

Solve the resulting linear equation in one variable.

$$\begin{aligned} 13x &= -39 &&\text{Divide each side by 13} \\ x &= -3 &&\text{We have our x!} \end{aligned}$$

Substitute $x = -3$ into one of the original equations to solve for y. Here, we will substitute $x = -3$ into the first equation, $-4x+5y = 12$.

$$\begin{aligned} -4(-3)+5y &= 12 &&\text{Simplify} \\ 12+5y &= 12 &&\text{Subtract 12 from each side} \\ 5y &= 0 &&\text{Divide each side by 5} \\ y &= 0 &&\text{We have our y!} \\ (-3,0) &&&\text{Our Solution} \end{aligned}$$

Verify that we have the correct solution by substituting $(x,y) = (-3,0)$ into each of the original equations.

First Equation:

$$-4x+5y = 12$$
$$-4(-3)+5(0) \stackrel{?}{=} 12$$
$$12+0 = 12 \checkmark$$

Second Equation:

$$-5x+3y = 15$$
$$-5(-3)+3(0) \stackrel{?}{=} 15$$
$$15+0 = 15 \checkmark$$

So the point $(-3,0)$ is a solution to the system of linear equations.

Exercise 5 Class Example

Solve the following system of linear equations and check your answer: $\begin{cases} 2g - 5h = 5 \\ 3g - 8h = 7 \end{cases}$

Exercise 6 You Try

Solve the following system of linear equations and check your answer: $\begin{cases} 5a - 4b = 1 \\ 11a + 6b = 17 \end{cases}$

3.4 Elimination Method

Just as with graphing and substitution, it is possible with the elimination method to have no solution or infinitely many solutions. If all the variables disappear from our problem, a true statement will indicate that we have Infinitely Many Solutions and a false statement will indicate that we have No Solution. It is important that the two linear equations are written in the same form before we begin the process of elimination.

Example 4

Solve the following system of linear equations: $\begin{cases} 2x = 5y + 3 \\ -6x + 15y = -9 \end{cases}$

Solution.
Notice that the two linear equations are not written in the same form. First subtract $5y$ from each side of the first equation.

$$\begin{aligned} 2x &= 5y+3 &\Rightarrow& \quad 2x - 5y &= 3 \\ -6x + 15y &= -9 &\Rightarrow& \quad -6x + 15y &= -9 \end{aligned}$$

We choose to eliminate the variable, x. The LCM of the coefficients of x is 6. Multiply the first equation by 3.

$$\begin{aligned} 3(2x - 5y) &= 3(3) &\Rightarrow& \quad 6x - 15y &= 9 \\ -6x + 15y &= -9 &\Rightarrow& \quad -6x + 15y &= -9 \\ \hline & & & \quad 0 &= 0 \end{aligned}$$

All the variables have been eliminated and we have a true statement. There are Infinitely Many Solutions.

Exercise 7 Class Example

Solve the following system of linear equations: $\begin{cases} 0.2x - 0.5y = -1 \\ -0.04x + 0.1y = 0.2 \end{cases}$

Exercise 8 You Try
Solve the following system of linear equations.

a) $\begin{cases} 2x - 3y = -4 \\ 0.6x - 0.9y = 1.2 \end{cases}$

b) $\begin{cases} 3x - y = 3 \\ 6x = 6 + 2y \end{cases}$

We have covered three different methods that can be used to solve a system of two linear equations with two variables. While all three can be used to solve any system of linear equations, each method has its own strengths. It is important that you are familiar with all three methods.

3.4: Exercises

Solve the following system by elimination

1. $9x + y = 22$
   $7x - y = 10$

2. $4m + 9n = -28$
   $-4m - 2n = 0$

3. $4a - 6b = -10$
   $4a - 6b = -14$

4. $2c - d = 5$
   $5c + 2d = -28$

5. $u - 2v = 5$
   $5u - 6v = 17$

6. $x + 3y = -1$
   $10x + 6y = -10$

7. $-\frac{2}{3}x + y = \frac{1}{3}$
   $\frac{1}{3}x - \frac{1}{2}y = \frac{1}{2}$

8. $-6x + 4y = 4$
   $-3x - y = 26$

9. $\frac{1}{4}u + \frac{1}{2}v = \frac{3}{2}$
   $\frac{1}{6}u + \frac{1}{3}v = 1$

10. $3a + 7b = -8$
    $2a + 3b = -2$

11. $-7c + 10d = 13$
    $4c + 9d = 22$

12. $g + 2h = 3$
    $-5g + 4h = -8$

13. $6a + 3b = -1$
    $8a + 9b = 2$

14. $-0.05x + 0.05y = 0.15$
    $0.1x - 0.1y = -0.3$

15. $0.1x + 0.06y = 0.24$
    $0.6x - 0.1y = -0.4$

16. $9x + 4y = -3$
    $3x + 12y = 7$

Solve each of the following by any method.

17. $x - y = -9$
    $y = -2x$

18. $9y = 7 - x$
    $9y + 2x = -13$

19. $9x - 2y = -12$
    $5x - 7y = 11$

20. $8x + 7y = -11$
    $y + 2 = -2x$

Mid-Chapter 3 Check-Up

Solve the following systems of linear equations by graphing.

1. $\begin{cases} y = \dfrac{2}{3}x \\ 2x - 3y = 6 \end{cases}$

2. $\begin{cases} x + 3y = 9 \\ 2x - y = 4 \end{cases}$

Solve the following systems of linear equations by substitution.

3. $\begin{cases} 3g + 4h = 2 \\ g = h + 3 \end{cases}$

4. $\begin{cases} d = 3c - 4 \\ d = 1 - 2c \end{cases}$

Solve the following systems of linear equations by elimination.

5. $\begin{cases} 2a - b = -4 \\ a + b = 1 \end{cases}$

6. $\begin{cases} 3m + 4n = -8 \\ 2m - 5n = 10 \end{cases}$

3.5 Applications of Systems of Linear Equations

Objective: To solve application problems using system of linear equations

Problem Solving Strategies and Tools (PSST)

When first looking at an application problem (or story problem), it is helpful to read the entire problem and then read it again more slowly to organize your thoughts. Note if additional information is needed.

1. Identify any unknown quantities and select variables to represent them.

2. Write the linear equations that model the relationship between the known and unknown quantities.

3. Solve the system of linear equations. Check for reasonableness of solution.

4. Report the solution in a complete sentence.

Mixture Problems

Example 1 A chemist has 70 mL of a 50% methane solution.

a) How much of an 80% methane solution must be added so the final solution is 60% methane?

b) How much final solution will the chemist have?

Solution.

Let us go through each step of the Problem Solving Strategies and Tools.

1. *Identify any unknown quantities and select variables to represent them.*
   Let x = amount of 80% methane solution
   Let y = amount of 60% methane solution

2. *Write the linear equations that model the relationship between the known and unknown quantities.*
   It is always a good idea to draw a picture to represent the situation.

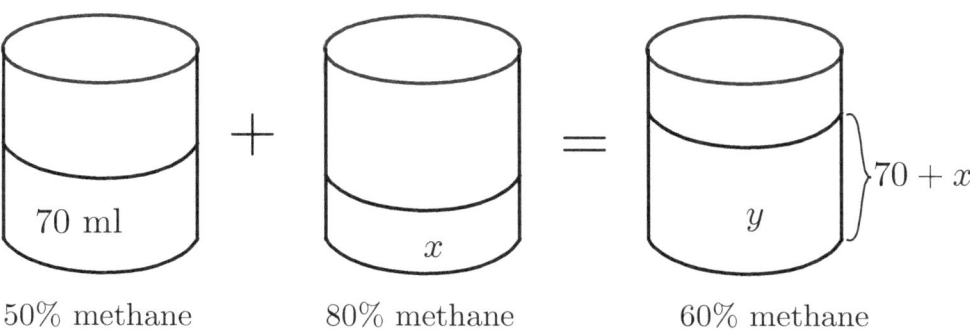

Set up the linear equations.

Equation 1: $70 + x = y$

(amount of 50% methane solution + amount of 80% methane solution = amount of final solution)

Equation 2: $50\%(70\text{mL}) + 80\%(x) = 60\%(y)$ or

$0.50(70) + 0.80x = 0.60y$

The system of linear equations is $\begin{cases} 70 + x = y \\ 0.50(70) + 0.80x = 0.60y \end{cases}$

3 *Solve the system of linear equations. Check for reasonableness of solution.*
In the first equation, we see that the variable y is expressed in terms of the variable x. We will use substitution to solve the system of linear equation. Substitute the first equation into the second equation, $0.50(70) + 0.80x = 0.60y$.

$0.50(70) + 0.80x = 0.60(70 + x)$	Distribute
$35 + 0.80x = 42 + 0.60x$	Subtract 35 from each side
$0.80x = 7 + 0.60x$	subtract 0.60x from each side
$0.20x = 7$	Divide each side by 0.20
$x = 35$	

Substitute $x = 35$ into the first equation and solve for y.

$70 + x = y$	Substitute $x = 35$
$70 + 35 = y$	Add
$105 = y$	

4 *Report the solution in a complete sentence.*
The chemist needs 35 mL of the 80% methane solution. The chemist will have 105 mL of the final solution.

3.5 Applications of Systems of Linear Equations

Exercise 1 Class Example
How many ounces of pure acid must be added to 40 ounces of a 20% acid solution to make a mixture that is 36% acid solution?

Exercise 2 You try
A solution of pure antifreeze is mixed with water to make a 65% antifreeze solution. How much of each should be used to make 70 L of the 65% antifreeze solution?

Example 2 Blondie's Candy Shop has gourmet chocolate which sells for $12.00 a pound and fancy nuts which sell for $7.50 a pound. Blondie would like to sell a chocolate and nut mix for $10.50 a pound. How much of each should she use to make 30 pounds of the new mixture?

Solution.
Let us go through each step of the Problem Solving Strategies and Tools.

1 *Identify ay unknown variables and select variables to represent them*
   Let c = amount of gourmet chocolate
   Let n = amount of fancy nuts

2 *Write the linear equations that model the relationship between the known and unknown quantities.*
   It is always a good idea to draw a picture to represent the situation.

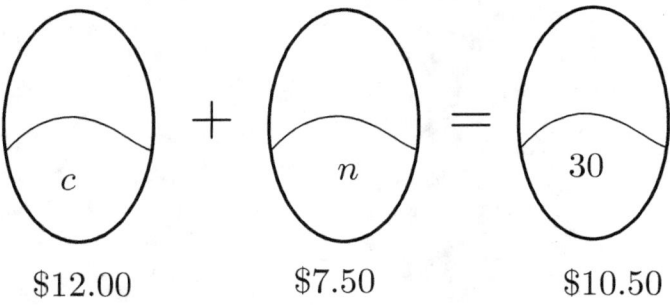

$\quad\quad\quad\quad$ \$12.00 $\quad\quad\quad\quad$ \$7.50 $\quad\quad\quad\quad$ \$10.50

Let us first set up the equations using words.
Equation 1: amount of gourmet chocolate + amount of fancy nuts = amount of chocolate and nut mixture
Equation 2: \$12×(amount of gourmet chocolate) + \$7.50×(amount of fancy nuts) = 30 lbs mixture at \$10.50/lb

Now let us set up the linear equations.
Equation 1: $c + n = 30$
Equation 2: $\$12.00(c) + \$7.50(n) = \$10.50(30)$

The system of linear equations is $\begin{cases} c + n = 30 \\ 12c + 7.5n = 315 \end{cases}$

3 *Solve the system of linear equations. Check for reasonableness of solution.*
   We will use elimination method to solve the system of linear equations. We choose to eliminate the variable, c. The LCM of the coefficients of c is 12. Multiply the first equation by -12.

$$\begin{aligned} -12(c+n) &= -12(30) \\ 12c + 7.5n &= 315 \end{aligned} \Rightarrow \begin{aligned} -12c - 12n &= -360 \\ 12c + 7.5n &= 315 \\ \hline -4.5n &= -45 \end{aligned}$$

3.5 Applications of Systems of Linear Equations

Solve the resulting linear equation in one variable.

$$-4.5n = -45 \qquad \text{Divide each side by } -4.5$$
$$n = 10$$

Substitute $n = 10$ into the first equation, $c + n = 30$ and solve for c.

$$c + n = 30 \qquad \text{Substitute } n = 10$$
$$c + 10 = 30 \qquad \text{Subtract 10 from each side}$$
$$c = 20$$

4 *Report the solution in a complete sentence.*
Blondie's Candy Shop needs 20 pounds of gourmet chocolate and 10 pounds of fancy nuts to make 30 pounds of chocolate and nut mixture that sells for $10.50 a pound.

Exercise 3 Class Example

The NorthStar Espresso stand wants to offer a special blend of coffee beans. They sell Kona beans for $9.15 per pound and Colombian beans for $8.75 per pound. How many pounds of each type of beans should the barista combine to create 50 pounds of a blend that can be sold for $9.00 per pound?

Exercise 4 Class Example
After the baseball game, the teams go to *Sammy's Sweet Shop*. The *Frogs* order 6 waffle cones and 6 banana splits for $55.50. The *Monkeys* order 3 waffle cones and 9 banana splits for $57.75. How much does each item cost?

Exercise 5 You Try
After their visit to the aquarium, the Mallards have lunch at a seafood bar. Fish and chips are $8.89 per order and a cup of chowder is $3.99. They purchase twice as many orders of fish and chips as cups of chowder and spent $87.08. How many orders of each did the Mallards buy?

3.5 Applications of Systems of Linear Equations

Distance, Rate and Time Problems

For this part, we will use the formula: Distance = Rate x Time.

Example 3 Mark and Jack start from the same point and run in opposite directions. Mark runs 2 miles per hour faster than Jack. After 3 hours, they are 30 miles apart. How fast did each run?

Solution.
Let us go through each step of the Problem Solving Strategies and Tools.

1 *Identify ay unknown variables and select variables to represent them*
   Let m = Mark's speed in miles per hour
   Let j = Jack's speed in miles per hour

2 *Write the linear equations that model the relationship between the known and unknown quantities.*
   Setting up a table is often useful for distance, rate and time problems. Note that distance = rate x time or $d = rt$.

Person	Distance	Rate	Time
Mark	$3m$	m	3
Jack	$3j$	j	3

   Draw a picture to represent the situation.

   Mark's distance $\longleftarrow\underbrace{ \circ }_{30 \text{ miles}}\longrightarrow$ Jack's distance

3 *Set up the linear equations.*
   Equation 1: $3m + 3j = 30$ (Mark's distance + Jack's distance = 30 miles)
   Equation 2: $m = j + 2$ (Mark runs 2 miles per hour faster than Jack

   The system of linear equations is $\begin{cases} 3m + 3j = 30 \\ m = j + 2 \end{cases}$

4 *Solve the system of linear equations. Check for reasonableness of solution.*
   Since the variable m is expressed in terms of variable j in the second equation, $m = j + 2$, we will use substitution to solve the system of linear equations. Substitute the second equation

into the first equation, $3m + 3j = 30$.

$$3m + 3j = 30 \qquad \text{Substitute } m = j + 2$$
$$3(j + 2) + 3j = 30 \qquad \text{Distribute the 3}$$
$$3j + 6 + 3j = 30 \qquad \text{Combine like terms}$$
$$6j + 6 = 30 \qquad \text{Subtract 6 from each side}$$
$$6j = 24 \qquad \text{Divide each side by 6}$$
$$j = 4$$

Substitute $j = 4$ into the second equation, $m = j + 2$ and solve for m.

$$m = j + 2 \qquad \text{Substitute } j = 4$$
$$m = 4 + 2 \qquad \text{Add}$$
$$m = 6$$

5 *Report the solution in a complete sentence.*
Mark ran at a rate of 6 miles per hour and Jack ran at a rate of 4 miles per hour.

Exercise 6 Class Example
Joey leaves his home for school, walking at a rate of 220 feet per minute. 15 minutes later, his mom realizes that Joey forgot his lunch. She gets on her bike, going at a rate of 1050 feet per minute. How long will it take Joey's mom to catch up with Joey? (Round answer to the nearest minute)

3.5 Applications of Systems of Linear Equations

Exercise 7 You Try

On a 130 mile family road trip to a mountain cabin, the Mallard's car traveled at an average speed of 60 mph on the highway and then reduced its speed to 40 mph when they drove over the snowy mountain pass. The trip took 2.5 hours. How long did the car travel at each speed?

3.5: Exercises

Solve the following.

1. Mary and Bobby are 4-year old twins. Their combined weight is 75 lbs and their weight difference is 7 lbs (Bobby being the heavier one). How much do each of them weigh?

2. The day before a predicted snowstorm, a hardware store sold five times more flashlights than snow shovels. The store sold 138 of both items that day. How many of each were sold that day?

3. Mrs. Smith is bringing dessert for the holiday party. Her baking time is 15 minutes longer than her prep time. It will take 65 minutes for the dessert to be ready. How long is the prep time and how long is the bake time?

4. How much pure alcohol must be added to 24 gallons of a 14% solution of alcohol in order to produce a 20% solution?

5. Mrs. Twinkle has tea that costs 26¢ per ounce that she will mix with 12 ounces of tea that cost 14¢ per ounce to form a new tea mix that will sell for 21¢ per ounce. How much 26¢ per ounce tea will Mrs. Twinkle use? How much will the new mixture weigh?

6. Blondie's Candy Shop has root beer barrels which sell for $15.00 per kilogram and butterscotch candies which sell for $6.50 per kilogram. Blondie would like to use 24 kilograms of butterscotch candies in the mix which she will sell for $9.00 per kilogram. How many kilograms of root beer barrels will she use? How much does the new mixture weigh?

7. A veterinarian told Barney that he needs to feed his dog with dog food that contains 11% fat. At the pet store, he can only find dog food with either 8% or 20% fat. Barney decides to mix his own dog food. If he wants 48 pounds of dog food, how many pounds of each pet store dog food does he need to buy?

8. Samantha's Syrup company manufactures some pure maple syrup and some syrup which is 85% maple syrup. How many quarts of each should she mix to make 150 quarts which is 96% maple syrup?

9. Two families visit the aquarium. The Smith family purchased 3 adult tickets and 2 youth tickets for $100.75. The Smiley family purchased an adult ticket and 4 youth tickets for $88.75. How much is one adult ticket and how much is one youth ticket?

10. Lily has exactly $5.06 to spend on trail mix for the algebra class hike and picnic. Chocolate candies cost 11¢ per ounce. Peanuts and raisins mix costs 22¢ per ounce. Lily wants 8 more ounces of peanuts and raisins mix than chocolate candies. How much of each will she be able to buy?

11. Wilma is making a platter for the algebra class picnic. A pound of ham costs twice as much as the pound of cheese. Wilma spends $48.75 on the platter and buys 5 pounds of each. How much is a pound of cheese and how much is a pound of ham?

12. There were 212 people at a play. Admission was $2.25 for each adult and 75¢ for each child. The total revenue from tickets was $283.50. How many children and how many adults attended?

3.5 Applications of Systems of Linear Equations

13. The algebra students spend $90.44 to grill burgers and veggie dogs at their picnic. Each burger costs $2.27 and each veggie dog costs $1.92. The students want to grill twice as many burgers as veggie dogs. How many of each will they grill?

14. A passenger train and a freight train, on parallel tracks, start towards the same train station from two points 300 miles apart. If the rate of the passenger train exceeds the rate of the freight train by 15 miles per hour and they meet after 4 hours, what is the rate of each train?

15. Howard walks and jogs to campus each day. He averages 5km/hr (kph) walking and 9 kph jogging. The distance from home to campus is 8 km and Howard makes the trip in one hour. How long does he jog and how long does he walk?

16. As part of his flight training, Luke was required to fly to an airport and then return following the same route. The average speed to the airport was 90 mph and the average speed returning was 120 mph. The total flying time was 7 hours.

    (a) Find how long it took to fly to the airport and how long it took to fly back.

    (b) Find the distance between the two airports.

17. Mark wants to jog and Jackie wants to rollerblade around Greenlake, a 2.8 mile distance. Mark's average jogging speed is 4 mph slower than Jackie's average rollerblading speed. If both of them go in opposite direction starting at the community center and they first pass each other 20 minutes (or $\frac{1}{3}$ hour) after they began, find each of their speeds.

18. 15 minutes (or $\frac{1}{4}$ hour) after Tommy got on the school bus, his mom realizes that Tommy forgot to bring his math homework. His mom gets in the car and drives at an average speed of 35 mph to try and catch up with the school bus. The school bus is going at an average speed of 20 mph. How long does it take for Tommy's mom to catch up to the school bus? (Round your answer to the nearest whole minute.)

Chapter 3 Assessment

Solve each system of linear equations by the indicated method.

1. By graphing: $\begin{cases} x+y=6 \\ y=2x \end{cases}$

2. By substitution: $\begin{cases} x=5y-3 \\ 2x-3y=8 \end{cases}$

3. By elimination: $\begin{cases} 3x+5y=-4 \\ 4x-3y=-15 \end{cases}$

Solve each system of linear equations by any method.

4. $\begin{cases} 9m-3n=12 \\ n=3m-4 \end{cases}$

5. $\begin{cases} 5a+2b=3 \\ a-b=2 \end{cases}$

6. $\begin{cases} 4y=3x+4 \\ 6x-8y=4 \end{cases}$

7. $\begin{cases} \frac{1}{2}x-\frac{1}{4}y=\frac{3}{2} \\ \frac{1}{3}x+\frac{1}{4}y=\frac{1}{6} \end{cases}$

Solve the following word problems.

8. At one elementary school, there are twice as many girls as boys in the third grade. There are 108 third graders. How many boys and how many girls are there?

9. A farmer wants to mix some 10% acid solution with some 30% acid solution to get 40 gallons of a 15% acid solution. How many gallons of 10% acid solution and how many gallons of 30% acid solutions should be mixed?

10. Robin went to see her grandmother who lives 69 miles away. She rode the bus traveling at 45 mph and then walked at 3 mph. The whole trip took 2 hours. How long did Robin walk?

4. Exponents

4.1 Introduction to Exponents

Objective: To introduce the concept of base and exponent

Given the expression a^n, a is called the **base** and n is called the **exponent**.

Positive Exponents

If the exponent of the expression is a positive integer, then a^n tells us that the base, a, is multiplied n times, that is,

$$a^n = \underbrace{a \cdot a \cdot a \cdots a}_{n \text{ times}}$$

> **World View Note.** French mathematician, Rene Descartes, popularized the superscript notation for showing powers or exponents. That is, instead of $2 \cdot 2 \cdot 2 \cdot 2 \cdot 2 \cdot 2$, Descartes popularized 2^6.

Example 1 Given the expression 3^4. Identify the base and the exponent. Then evaluate the expression.

Solution.

The base is 3 and the exponent is 4. That means, multiply the base, 3, four times or $3^4 = 3 \cdot 3 \cdot 3 \cdot 3 = 81$.

Exercise 1 Class Example
Given the expression 2^5. Identify the base and the exponent. Then evaluate the expression.

Exercise 2 You Try
Given the expression 4^2. Identify the base and the exponent. Then evaluate the expression.

To simplify the expression correctly, it is very important to be able to identify the base and its exponent accurately. The order of operations must also be followed.

Example 2 Simplify $4 \cdot 3^2$ and $(4 \cdot 3)^2$.

Solution.
Let's take a look at the expression $4 \cdot 3^2$.

$4 \cdot 3^2 = 4 \cdot 9$ Exponent first
$ = 36$ Our Solution

Another way to look at this problem is to determine what is the base for the exponent 2. In this expression, $4 \cdot 3^2$, only 3 is the base for the exponent 2. The expression can be rewritten as follows.

$4 \cdot 3^2 = 4 \cdot 3 \cdot 3$
$ = 12 \cdot 3$
$ = 36$

Now let's look at the expression $(4 \cdot 3)^2$.

$(4 \cdot 3)^2 = (12)^2$ Parenthesis first
$ = 144$ Our Solution

In this expression, $(4 \cdot 3)^2$, the product $(4 \cdot 3)$ is the base for the exponent 2. The expression can be rewritten as follows.

$(4 \cdot 3^2) = (4 \cdot 3)(4 \cdot 3)$
$ = (12)(12)$
$ = 144$

4.1 Introduction to Exponents

Example 3 Expand $4x^3$ and $(4x)^3$.

Solution.
In the expression, $4x^3$, base 4 has an exponent 1 and base x has exponent 3. Therefore,

$$4x^3 = 4 \cdot x \cdot x \cdot x$$

In the expression, $(4x)^3$, the base is $(4x)$ and it has an exponent 3.

$$(4x)^3 = (4x)(4x)(4x)$$
$$= 64x^3$$

Exercise 3 Class Example
Expand the following expressions.

a) $-y^4$ and $(-y)^4$

b) $\dfrac{p^5}{2}$ and $\left(\dfrac{p}{2}\right)^5$

Exercise 4 You Try
Expand the following expressions.

a) $(-3)^4$ and -3^4

c) $5 \cdot 2^3$ and $(5 \cdot 2)^3$

b) $\dfrac{5^2}{6}$ and $\left(\dfrac{5}{6}\right)^2$

d) $(-x)^2$ and $-x^2$

4.1: Exercises

Expand the following expressions.

1. w^4
2. $(-w)^2$
3. $-w^2$
4. $(2w)^4$
5. $2w^4$
6. $\left(\dfrac{5}{p}\right)^3$
7. $\dfrac{5^3}{p}$
8. $\dfrac{5^3}{p^3}$
9. $\dfrac{5}{p^3}$
10. $(x-7)^2$
11. $x^2 - 7^2$
12. g^3
13. $-g^3$
14. $(-g)^3$
15. $7g^5$
16. $(7g)^5$
17. $\dfrac{g}{-7^5}$
18. $\dfrac{g^5}{7}$
19. $\left(\dfrac{g}{-7}\right)^5$
20. $(k+2)^4$
21. $k^4 + 2^4$

4.2 Properties of Exponents

Objective: To simplify expressions using the properties of exponents

Expressions with exponents can often be simplified using a few basic exponent properties. Exponents represent repeated multiplication. We will use this fact to discover the important properties.

Product Rule

Example 1 Simplify $a^3 a^2$

Solution

$a^3 a^2 = (a \cdot a \cdot a) \cdot (a \cdot a)$ Expand
$ = a^5$ Our Solution

From the above example, we see that the outcome is the same if we added the exponents. That is, $a^3 \cdot a^2 = a^5$. When multiplying two expressions that have the same base, we can add exponents. This is known as the **product rule of exponents.**

Product Rule of Exponents:

$$a^m a^n = a^{m+n}$$

Example 2 Simplify $3^6 \cdot 3$

Solution

$3^6 \cdot 3 = (3 \cdot 3 \cdot 3 \cdot 3 \cdot 3 \cdot 3) \cdot 3$ Expand
$ = 3^7$ Our Solution

Example 3 Simplify $(2x^3) \cdot (5x)$

Solution

$(2x^3) \cdot (5x) = (2 \cdot x \cdot x \cdot x) \cdot (5 \cdot x)$ Expand
$ = 10x^4$ Our Solution

Exercise 1 Class Example
Simplify the following.

a) $p^5 p^7$

b) $(3m^4)(4m^2)$

Exercise 2 You Try
Simplify the following.

a) $x^8 x^2$
b) $(5x^2)(3x)$

Quotient Rule

In division, common factors from the numerator and denominator simplify to a one.

Example 4 Simplify $\dfrac{a^5}{a^2}$

Solution

$\dfrac{a^5}{a^2} = \dfrac{a \cdot a \cdot a \cdot a \cdot a}{a \cdot a}$ Expand

$= \dfrac{a}{a} \cdot \dfrac{a}{a} \cdot a \cdot a \cdot a$ Divide common factors

$= 1 \cdot 1 \cdot a \cdot a \cdot a$ Note that $\dfrac{a}{a} = 1$

$= a^3$ Our solution

From the above example, we see that the outcome is the same if we subtracted exponents. That is, $\dfrac{a^5}{a^2} = a^{5-2} = a^3$. When dividing two expressions that have the same base, we can subtract exponents as long as the base is not equal to zero. This is known as the **quotient rule of exponents.**

Quotient Rule of Exponents:

$$\dfrac{a^m}{a^n} = a^{m-n}, \quad a \neq 0$$

4.2 Properties of Exponents

Example 5 Simplify $\dfrac{c^3}{c^5}$

Solution

$$\dfrac{c^3}{c^5} = \dfrac{c \cdot c \cdot c}{c \cdot c \cdot c \cdot c \cdot c} \qquad \text{Expand}$$

$$= \dfrac{c}{c} \cdot \dfrac{c}{c} \cdot \dfrac{c}{c} \cdot \dfrac{1}{c \cdot c} \qquad \text{Divide common factors}$$

$$= 1 \cdot 1 \cdot 1 \cdot \dfrac{1}{c \cdot c} \qquad \text{Note that } \dfrac{c}{c} = 1$$

$$= \dfrac{1}{c^2} \qquad \text{Our solution}$$

Example 6 Simplify $\dfrac{7^{10}}{7^4}$

Solution

$$\dfrac{7^{10}}{7^4} = \dfrac{7 \cdot 7 \cdot 7 \cdot 7 \cdot 7 \cdot 7 \cdot 7 \cdot 7 \cdot 7 \cdot 7}{7 \cdot 7 \cdot 7 \cdot 7} \qquad \text{Expand and simplify}$$

$$= 7 \cdot 7 \cdot 7 \cdot 7 \cdot 7 \cdot 7 \qquad \text{Note that } \dfrac{7}{7} = 1$$

$$= 7^6 \qquad \text{Our solution}$$

Example 7 Simplify $\dfrac{15a^3}{3a}$

Solution

$$\dfrac{15a^3}{3a} = \dfrac{5 \cdot 3 \cdot a \cdot a \cdot a}{3 \cdot a} \qquad \text{Expand and simplify}$$

$$= 5 \cdot a \cdot a \qquad \text{Note that } \dfrac{3}{3} = 1 \text{ and } \dfrac{a}{a} = 1$$

$$= 5a^2 \qquad \text{Our solution}$$

Exercise 3 Class Example
Simplify the following.

a) $\dfrac{p^7}{p^4}$

b) $\dfrac{10n^6}{5n^6}$

Exercise 4 You Try
Simplify the following.

a) $\dfrac{x^9}{x^2}$

b) $\dfrac{12y^5}{8y^3}$

Power Rule of Exponents

Example 8 Simplify $(a^2)^3$

Solution

$(a^2)^3 = a^2 \cdot a^2 \cdot a^2$ — Multiply a^2 by itself three times
$= (a \cdot a) \cdot (a \cdot a) \cdot (a \cdot a)$ — a^2 expanded
$= a \cdot a \cdot a \cdot a \cdot a \cdot a$ — Rewrite without parenthesis
$= a^6$ — Our solution

From the above example, we see that the outcome is the same if we multiply the exponents. That is, $(a^2)^3 = a^{2 \cdot 3} = a^6$. When a base with an exponent is raised to an exponent, we can multiply the two exponents, keeping the base. This is known as the **power rule of exponents.**

Power Rule of Exponents:

$$(a^m)^n = a^{mn}$$

Example 9 Simplify $(7^4)^3$

Solution

$(7^4)^3 = 7^4 \cdot 7^4 \cdot 7^4$ — Multiply 7^4 by itself three times
$= (7 \cdot 7 \cdot 7 \cdot 7) \cdot (7 \cdot 7 \cdot 7 \cdot 7) \cdot (7 \cdot 7 \cdot 7 \cdot 7)$ — 7^4 expanded
$= 7 \cdot 7 \cdot 7 \cdot 7 \cdot 7 \cdot 7 \cdot 7 \cdot 7 \cdot 7 \cdot 7 \cdot 7 \cdot 7$ — Rewrite without parenthesis
$= 7^{12}$ — Our solution

Exercise 5 Class Example
Simplify the following.

a) $(m^3)^4$

b) $(9^7)^3$

4.2 Properties of Exponents

Exercise 6 You Try

Simplify the following.

a) $(p^4)^5$

b) $(5^{10})^2$

Power of a Product Rule

Example 10 Simplify $(ab)^3$

Solution

$$(ab)^3 = (ab) \cdot (ab) \cdot (ab) \quad \text{Expand}$$
$$= a \cdot a \cdot a \cdot b \cdot b \cdot b \quad \text{Rewrite without parenthesis and apply commutative property}$$
$$= a^3 b^3 \quad \text{Our solution}$$

From the above example, we see that when different factors are raised to an exponent, each factor is raised to that power. This is known as the **power of a product rule.**

Power of a Product Rule:

$$(ab)^m = a^m b^m$$

It is important to only use the power of a product rule with factors, meaning, multiplication inside the parenthesis. A common error is to use the rule for a *sum* or a *difference* raised to a power. The rule does not apply to sum or difference.

Warning. $(a \pm b)^m \neq a^m \pm b^m$. **Beware of this error!!**

Example 11 Simplify $(5m)^2$

Solution

$$(5m)^2 = 5m \cdot 5m \quad \text{Multiply 5m by itself}$$
$$= 25m^2 \quad \text{Our solution}$$

Exercise 7 Class Example
Simplify the following.

a) $(ab)^2$

b) $(4c)^3$

Exercise 8 You Try
Simplify the following.

a) $(mn)^4$

b) $(2a)^5$

Power of a Quotient Rule

Example 12 Simplify $\left(\dfrac{a}{b}\right)^3$

Solution

$\left(\dfrac{a}{b}\right)^3 = \left(\dfrac{a}{b}\right) \cdot \left(\dfrac{a}{b}\right) \cdot \left(\dfrac{a}{b}\right)$ Multiply $\left(\dfrac{a}{b}\right)$ three times

$= \dfrac{a^3}{b^3}$ Our solution

From the above example, we see that when a quotient is raised to an exponent, the numerator and denominator are raised to that power, as long as the denominator is not equal to zero. This is known as the **power of a quotient rule**.

Power of a Quotient Rule:

$$\left(\dfrac{a}{b}\right)^m = \dfrac{a^m}{b^m}, \quad b \neq 0$$

4.2 Properties of Exponents

Example 13 Simplify $\left(\dfrac{7}{m}\right)^2$

Solution

$\left(\dfrac{7}{m}\right)^2 = \left(\dfrac{7}{m}\right) \cdot \left(\dfrac{7}{m}\right)$ Multiply $\left(\dfrac{7}{m}\right)$ by itself

$= \dfrac{49}{m^2}$ Our solution

Exercise 9 Class Example
Simplify the following.

a) $\left(\dfrac{x}{y}\right)^5$

b) $\left(\dfrac{2}{m}\right)^4$

Exercise 10 You Try
Simplify the following.

a) $\left(\dfrac{a}{b}\right)^7$

b) $\left(\dfrac{w}{6}\right)^2$

The product rule, quotient rule, and power of a power are often used together to simplify expressions. There can be a bit of flexibility as to which property to use first. However, you must always follow the order of operations.

Example 14 Simplify $(x^3 y^2)^4$

Solution

$(x^3 y^2)^4 = (x^3)^4 (y^2)^4$
$= x^{12} y^8$ Our solution

Example 15 Simplify $\left(\dfrac{a^3}{c^8}\right)^2$

Solution

$$\left(\frac{a^3}{c^8}\right)^2 = \frac{(a^3)^2}{(c^8)^2}$$ 	Use Power of a Product Rule

$$= \frac{a^6}{c^{16}}$$ 	Our solution

Example 16 Simplify $(4x^2y^5)^3$

Solution

$$(4x^2y^5)^3 = (4)^3(x^2)^3(y^5)^3$$ 	Use Power of a Product Rule
$$= 64x^6y^{15}$$ 	Our solution

Exercise 11 Class Example
Simplify the following.

a) $(7h^5)^2$

c) $4m^2(3m^4)^2$

b) $\left(\frac{2m^3}{p^5}\right)^4$

d) $\dfrac{12a^5}{(2a^2)^3}$

4.2 Properties of Exponents

Exercise 12 You Try
Simplify the following.

a) $(2w^7)^3$

b) $\left(\dfrac{w^3}{6y}\right)^2$

c) $9x^3(x^2)^4$

d) $\dfrac{(5c^3)^2}{10c^6}$

In this section, we have discussed five different exponent rules. These rules are summarized in the following table.

Rules of Exponents		Example
Product Rule	$a^m a^n = a^{m+n}$	$a^5 \cdot a^2 = a^7$
Quotient Rule	$\dfrac{a^m}{a^n} = a^{m-n}, a \neq 0$	$\dfrac{a^5}{a^2} = a^3$
Power Rule	$(a^m)^n = a^{mn}$	$(a^5)^2 = a^{10}$
Power of a Product Rule	$(ab)^m = a^m b^m$	$(ab)^5 = a^5 b^5$
Power of a Quotient Rule	$\left(\dfrac{a}{b}\right)^m = \dfrac{a^m}{b^m}, b \neq 0$	$\left(\dfrac{a}{b}\right)^5 = \dfrac{a^5}{b^5}$

4.2: Exercises

Simplify.

1. $9 \cdot 9^4 \cdot 9^2$
2. $x^2 \cdot x^3 \cdot x$
3. $3x \cdot 4x^2$
4. $(3m)(4mn)$
5. $(2m^4n^2)(4nm^2)$
6. $x^2y^4 \cdot xy^2$
7. $(7^2)^4$
8. $(y^3)^4$
9. $(xy)^3$
10. $(2a^4)^4$
11. $(2xy)^4$
12. $(2u^3v^2)^2$
13. $\dfrac{p^5}{p^3}$
14. $\dfrac{3^2}{3}$
15. $\dfrac{8y^4}{4y}$
16. $\dfrac{3nm^2}{6n}$
17. $\dfrac{10x^2y^4}{5xy}$
18. $\dfrac{4x^3y^4}{3xy^3}$
19. $\dfrac{2xy^3}{4xy}$
20. $\dfrac{(3y^4)^2}{y^8}$
21. $\left(\dfrac{7m^{14}}{m^3}\right)^2$
22. $2x(x^4y^4)^4$
23. $\left(\dfrac{(2x)^3}{x^3}\right)^2$
24. $\dfrac{2a^2b^2}{(a^4b)^2}$
25. $\dfrac{x^2y \cdot (y^4)^2}{2y^4}$
26. $\dfrac{m^3(n^4)^2}{2mn}$
27. $(xy)^3$
28. $(2xy^3)^4$

4.3 Negative Exponents and Zero Power

Objective: To simplify expressions with negative exponents using the properties of exponents

Zero Power

Example 1 What do you notice about the expression's expanded form and value as its base remains the same but its exponent decreases by 1?

Expression	Expanded Form	Value
4^5	$4 \cdot 4 \cdot 4 \cdot 4 \cdot 4$	1024
4^4	$4 \cdot 4 \cdot 4 \cdot 4$	256
4^3	$4 \cdot 4 \cdot 4$	64
4^2	$4 \cdot 4$	16
4^1	4	4
4^0	1	1

Notice that as the expression's exponent decreases by one, it's value decreases by a multiple of 4, which is its base.

Exercise 1 Class Example
Complete the table. What do you notice?

Expression	Expanded Form	Value
2^5	$2 \cdot 2 \cdot 2 \cdot 2 \cdot 2$	32
2^4	$2 \cdot 2 \cdot 2 \cdot 2$	16
2^3		
2^2		
2^1		
2^0		

Exercise 2 You Try
Complete the table. What do you notice?

Expression	Expanded Form	Value
3^5	$3 \cdot 3 \cdot 3 \cdot 3 \cdot 3$	243
3^4	$3 \cdot 3 \cdot 3 \cdot 3$	81
3^3		
3^2		
3^1		
3^0		

This suggests the following definition.

> **Zero Property of Exponents:**
>
> $a^0 = 1$, where a is any real number such that $a \neq 0$

Example 2 Simplify the following.

a) 5^0

Solution.

$5^0 = 1$ by the zero property of exponents where the base is 5 raised to the power 0.

b) $(-5)^0$

Solution.

$(-5)^0 = 1$ by the zero property of exponents where base, -5, is raised to the power 0.

c) -5^0

Solution.

$-5^0 = -1$; Note that the exponent 0 applies only to base 5. -5^0 can be rewritten as $-(5^0)$. Therefore, $-5^0 = -(5^0) = -(1) = -1$

d) $-5x^0, x \neq 0$

Solution.

$-5x^0 = -5$; Exponent 0 applies only to the base x. That means, $x^0 = 1$. However, -5 is multiplied by x^0. Therefore, $-5x^0 = -5(x^0) = -5(1) = -5$

4.3 Negative Exponents and Zero Power

Exercise 3 Class Example
Simplify the following.

a) 3^0

b) -4^0

c) $(6c)^0$, $c \neq 0$

d) $\dfrac{7^0}{2}$

e) $\left(\dfrac{7}{2}\right)^0$

f) $5^0 + 5^2$

Exercise 4 You Try
Simplify the following.

a) 8^0

b) -8^0

c) $(-8)^0$

d) $3 \cdot 5^0$

e) $(3 \cdot 5)^0$

f) $2^0 - 2^3$

Negative Exponents

Let us revisit the table at the beginning of the section. This time, we will extend the table further to include negative exponents.

Exercise 5 Class Example
Complete the table. What do you notice?

Expression	Expanded Form	Value
2^5	$2 \cdot 2 \cdot 2 \cdot 2 \cdot 2$	32
2^4	$2 \cdot 2 \cdot 2 \cdot 2$	16
2^3	$2 \cdot 2 \cdot 2$	8
2^2		
2^1		
2^0		
2^{-1}		
2^{-2}		
2^{-3}		
2^{-4}		
2^{-5}		

4.3 Negative Exponents and Zero Power

Exercise 6 You Try

Complete the table. What do you notice?

Expression	Expanded Form	Value
3^5	$3 \cdot 3 \cdot 3 \cdot 3 \cdot 3$	243
3^4	$3 \cdot 3 \cdot 3 \cdot 3$	81
3^3	$3 \cdot 3 \cdot 3$	27
3^2		
3^1		
3^0		
3^{-1}		
3^{-2}		
3^{-3}		
3^{-4}		
3^{-5}		

Example 3 Notice what happens to the expression's expanded form and value as its base remains the same but its exponent decreases by 1.

Expression	Expanded Form	Value
4^5	$4 \cdot 4 \cdot 4 \cdot 4 \cdot 4$	1024
4^4	$4 \cdot 4 \cdot 4 \cdot 4$	256
4^3	$4 \cdot 4 \cdot 4$	64
4^2	$4 \cdot 4$	16
4^1	4	4
4^0	1	1
4^{-1}	$\dfrac{1}{4}$	$\dfrac{1}{4}$
4^{-2}	$\dfrac{1}{4 \cdot 4}$	$\dfrac{1}{16}$
4^{-3}	$\dfrac{1}{4 \cdot 4 \cdot 4}$	$\dfrac{1}{64}$
4^{-4}	$\dfrac{1}{4 \cdot 4 \cdot 4 \cdot 4}$	$\dfrac{1}{256}$
4^{-5}	$\dfrac{1}{4 \cdot 4 \cdot 4 \cdot 4 \cdot 4}$	$\dfrac{1}{1024}$

Each time the expression's exponent decreases by 1, the value is divided by its base, 4. Notice also that the value of the expression with a negative exponent is the reciprocal of the value of the same expression with a positive exponent.

4.3 Negative Exponents and Zero Power

This leads us to the following property of negative exponents.

> **Property of Negative Exponents**
>
> If a is any real number such that $a \neq 0$ and n a positive integer, then
>
> $$a^{-n} = \frac{1}{a^n}$$
>
> That is, the negative power of a base is the reciprocal of the positive power of that base.

Given a^n where $a \neq 0$, the reciprocal of a^n is $\frac{1}{a^n} = a^{-n}$. This means that if you take the reciprocal of a^n, the sign of its exponent changes. If you think of expressions as fractions, once the factor is moved either to the numerator or denominator, the sign of the exponent changes while the base remains the same. Remember that it is the sign of the exponent that changes and **not** the base.

Warning. Negative exponents **never** make the base negative!

Example 4 Simplify each of the following.

a) 5^{-2}

Solution.

$5^{-2} = \dfrac{1}{5^2}$ Property of negative exponents; base is 5

$\phantom{5^{-2}} = \dfrac{1}{25}$ Our solution

b) $(-5)^{-2}$

Solution.

$(-5)^{-2} = \dfrac{1}{(-5)^2}$ Property of negative exponents; base is -5

$\phantom{(-5)^{-2}} = \dfrac{1}{25}$ Our solution

c) -5^{-2}

Solution.

$$-5^{-2} = -\frac{1}{5^2}$$ Note that base of the exponent -2 is 5

$$= -\frac{1}{25}$$ Our solution

d) $\dfrac{1}{5^{-2}}$

Solution.

$$\frac{1}{5^{-2}} = 5^2$$ Property of negative exponents

$$= 25$$ Our solution

Exercise 7 Class Example
Simplify the following.

a) 4^{-2}

d) $\left(\dfrac{2}{5}\right)^{-3}$

b) -4^{-2}

e) $\dfrac{1}{5^{-3}}$

c) $(-4)^{-2}$

f) $4^0 - 4^{-2}$

4.3 Negative Exponents and Zero Power

Exercise 8 You Try
Simplify the following.

a) 2^{-4}

b) $(-2)^{-4}$

c) -2^{-5}

d) $\dfrac{3}{2^{-3}}$

e) $\left(\dfrac{1}{2}\right)^{-5}$

f) $2^{-2} + 2^{-3}$

Exercise 9 Class Example
Simplify the following. Write your answers with positive exponents only.

a) $\dfrac{1}{m^{-4}}$

b) $-5x^{-3}$

c) $\dfrac{7}{w^{-3}}$

d) $\dfrac{m^5}{m^{-3}}$

Exercise 10 You Try
Simplify the following. Write your answers with positive exponents only.

a) p^{-5}

b) $7w^{-2}$

c) $\dfrac{c^{-3}}{2}$

d) $-m^{-1}$

Sometimes, in order to simplify an expression containing exponents, we may use a combination of properties of exponents.

Example 5 Simplify each of the following. Write the answer with positive exponents only.

a) $\left(\dfrac{x^5}{x^{-3}}\right)^6$

Solution.

$\left(\dfrac{x^5}{x^{-3}}\right)^6 = \dfrac{(x^5)^6}{(x^{-3})^6}$ Apply power of a quotient rule

$= \dfrac{x^{30}}{x^{-18}}$ Apply Property of negative exponents

$= x^{30} x^{18}$ Apply product rule

$= x^{48}$ Our Solution

4.3 Negative Exponents and Zero Power

b) $(4x^{-5})^{-3}$

Solution.

$(4x^{-5})^{-3} = (4)^{-3}(x^{-5})^{-3}$ Apply power of a product rule

$= 4^{-3}x^{15}$ Apply Property of negative exponents

$= \dfrac{x^{15}}{4^3}$ Calculate 4^3

$= \dfrac{x^{15}}{64}$ Our solution

c) $\dfrac{-8m^3}{4m^{-2}}$

Solution.

$\dfrac{-8m^3}{4m^{-2}} = \dfrac{-8m^3 m^2}{4}$ Apply Property of negative exponents

$= -2m^5$ Our solution

Exercise 11 Class Example
Simplify each of the following. Write your answers with positive exponents only.

a) $\dfrac{12ab^0}{4a^9 b^{-5}}$

c) $\dfrac{-15x^5}{30x^4}$

b) $-\left(\dfrac{2}{m}\right)^{-3}$

d) $\left(\dfrac{w^5}{-2w^{-4}}\right)^{-3}$

Exercise 12 You Try
Simplify each of the following. Write your answers with positive exponents only.

a) $\dfrac{-3x^5}{3x^5}$

b) $p(2p)^{-5}$

c) $10^{14} \cdot 10^{-6}$

d) $\left(\dfrac{w^5}{-2w^{-4}}\right)^3$

World View Note. Nicolas Chuquet, the French mathematician of the 15th century, may have been the first mathematician to recognize zero and negative exponents.

4.3: Exercises

Simplify. Answers should contain only positive exponents.

1. $2 \cdot 8^0$
2. $\left(\dfrac{3}{5}\right)^0$
3. 5^2
4. -5^2
5. $(-5)^2$
6. 5^{-2}
7. 4^3
8. 4^{-3}
9. $(-4)^3$
10. -4^3
11. $8^0 - 2^{-3}$
12. $3^{-2} + \left(\dfrac{1}{3}\right)^0$
13. $19^0 - 6^{-1}$
14. $\dfrac{3}{5}(7)^0 + \left(\dfrac{73}{5}\right)^0$
15. -3^{-2}
16. $(-3)^{-2}$
17. $-x^{-4}$
18. $(-x)^{-4}$
19. $3w^{-5}$
20. $\dfrac{x^4}{x^{-16}}$

21. $\dfrac{r^{-4}}{r^{-10}\, r^5}$
22. $(y^5)^{-1}$
23. $\left(\dfrac{1}{4}\right)^{-1}$
24. $\left(\dfrac{5}{k^4}\right)^{-1}$
25. $\left(\dfrac{b^{-2}}{2}\right)^{-3}$
26. $\dfrac{4}{5}(10^{-3}p^7)^0$
27. $\dfrac{b^{15}b^{-7}}{b^3 b^5}$
28. $\left(\dfrac{-2n^{-4}}{n^2}\right)^3$
29. $\left(\dfrac{3y^5}{12y^3}\right)^{-2}$
30. $\dfrac{-20c^{-8}}{5c^3}$
31. $\dfrac{6(x^{-2})^3}{x^{-10}}$
32. $\left(\dfrac{2y^{-4}}{y^2}\right)^{-2}$
33. $\dfrac{(a^4)^2}{-2a^8}$
34. $\dfrac{(a^{-4})^2}{-2a^8}$

Insert the symbols $>$, $<$, or $=$ to make the statement true without using a calculator.

35. 5^{20} ☐ 5^{-20}

36. -5^{20} ☐ $(-5)^{20}$

37. -5^{20} ☐ $\dfrac{1}{5^{20}}$

38. $\left(\dfrac{1}{5}\right)^{20}$ ☐ 5^{-20}

39. $\dfrac{1^{20}}{5}$ ☐ $\dfrac{1}{5}$

40. 500^0 ☐ 5^0

41. $(-3000)^{24}$ ☐ -3000^{24}

42. 3000^0 ☐ 3000^{-1}

43. $(-3000)^0$ ☐ -30000^0

44. $\dfrac{1}{3000}$ ☐ 3000^{-1}

Rescue Roody!

Roody was given the following problems to simplify. His answers were all incorrect. Help Roody figure out how to correctly simplify the expressions.

45. Simplify $\dfrac{3^0}{4}$.

    Roody's answer: 1. Roody says that any base different from 0 and raised to the power 0 is 1.

46. Simplify 2^{-3}.

    Roody's answer: -6. Roody thinks that to simplify the expression, multiply the base 2 and exponent -3 to get the answer -6.

47. Simplify $-6x^{-3}$.

    Roody's answer: $\dfrac{1}{6x^3}$. Because 6 is negative and x has a negative exponent, Roody took the reciprocal of both factors and changed the negative sign to a positive sign.

4.4 Scientific Notation

Objective: To understand scientific notation and convert between standard notation and scientific notation

Numbers that are very large or very small can sometimes be difficult to understand or write in standard form. An example of a very large number is the distance that light travels in a year, approximately 6,000,000,000,000 miles. An example of a really small number is the mass of a hydrogen atom, approximately 0.00000000000000000000000166 grams. As you can see, these numbers are quite difficult to understand and tedious to write. Scientific notation was developed to make these very large or very small numbers easier to read, write and understand.

Numbers in scientific notation have the following form:

> **Scientific Notation**
> A number N written is Scientific Notation has the form
> $$N = a \times 10^b,$$
> where $1 \leq a < 10$ and b is an integer.

The coefficient, a, must be greater than or equal to 1, and strictly less than 10. The exponent of 10, b, tells us how many times we must multiply or divide by 10.

Example 1 Convert 4.3×10^3 to standard notation.

Solution.

$$\begin{aligned}
4.3 \times 10^3 &= 4.3 \times \underbrace{10 \cdot 10 \cdot 10} && \text{Expand } 10^3 \\
&= 4.3 \times 1000 && \text{Multiply} \\
&= 4300 && \text{Standard Notation}
\end{aligned}$$

In effect, we moved the decimal 3 places to the right, adding zeros as necessary. The positive exponent of base 10 indicates that the value of the entire number is larger than the coefficient, 4.3.

Exercise 1 **Class Example**
Convert 2×10^5 to standard notation.

How about when the exponent of base, 10, is negative?

Example 2 Convert 5.7×10^{-4} to standard notation.

Solution.

$5.7 \times 10^{-4} = 5.7 \times \dfrac{1}{10^4}$ Apply property of negative exponents

$\phantom{5.7 \times 10^{-4}} = 5.7 \times \dfrac{1}{10 \cdot 10 \cdot 10 \cdot 10}$ Expand 10^4

$\phantom{5.7 \times 10^{-4}} = 5.7 \times \dfrac{1}{10,000}$ Multiply factors of 10

$\phantom{5.7 \times 10^{-4}} = 5.7 \times 0.0001$ Multiply

$\phantom{5.7 \times 10^{-4}} = 0.00057$ Standard Notation

In effect, we moved the decimal 4 places to the left, introducing zeros as necessary. The negative exponent of base 10 indicates that the value of the entire number is smaller than the coefficient, 5.7.

Exercise 2 Class Example
Convert 6×10^{-1} to standard notation.

Exercise 3 You Try
Determine which value is larger and how many times larger than the other value.

a) 10^3 or 10^5

b) 10^{-2} or 10^{-4}

c) 10^0 or 10^{-1}

d) 10^{-3} or 10^3

4.4 Scientific Notation

Exercise 4 You Try
Convert each of the following from scientific notation to standard notation.

a) 3.42×10^6

b) 7.85×10^{-3}

c) 8×10^9

d) 1.9×10^{-4}

Now that we understand how to convert from scientific notation to standard notation, we will learn how to convert from standard notation to scientific notation.

Example 3 Convert 123,000 to scientific notation.

Solution.
Our goal is to rewrite the number in the form $a \times 10^b$ where $1 \leq a < 10$ and b is an integer. Since 123,000 is larger than 1, we will do this by factoring out 10 until the coefficient, a, is between 1 and 10.

$$123,000 = 12,300 \times 10$$
$$= 1,230 \times 10 \cdot 10$$
$$= 123 \times 10 \cdot 10 \cdot 10$$
$$= 12.3 \times 10 \cdot 10 \cdot 10 \cdot 10$$
$$= 1.23 \times 10 \cdot 10 \cdot 10 \cdot 10 \cdot 10$$
$$123,000 = 1.23 \times 10^5 \quad \text{(in scientific notation)}$$

In effect, we moved the decimal 5 places to the left. As the coefficient becomes smaller, the exponent of 10 must become larger for the number in standard and scientific notation to remain equivalent.

Example 4 Convert 0.0028 to scientific notation.

Solution.
Our goal is to rewrite the number in the form $a \times 10^b$ where $1 \leq a < 10$ and b is an integer. Since 0.0028 is smaller than 1, we will do this by factoring out $0.1 = \dfrac{1}{10} = 10^{-1}$ until the coefficient,

a, is between 1 and 10.

$$0.0028 = 0.028 \times 0.1 \qquad\qquad = 0.028 \times \frac{1}{10}$$
$$= 0.28 \times (0.1)(0.1) \qquad\qquad = 0.28 \times \frac{1}{10} \cdot \frac{1}{10}$$
$$= 2.8 \times (0.1)(0.1)(0.1) \qquad\qquad = 2.8 \times \frac{1}{10} \cdot \frac{1}{10} \cdot \frac{1}{10}$$
$$= 2.8 \times 0.001 \qquad\qquad = 2.8 \times \frac{1}{10^3}$$
$$0.0028 = 2.8 \times 10^{-3} \qquad\qquad \text{(in scientific notation)}$$

In effect, we moved the decimal 3 places to the right. As the coefficient becomes larger, the exponent of 10 must become smaller for the number in standard and scientific notation to remain equivalent.

Exercise 5 Class Example
Convert each of the following from standard notation to scientific notation.

a) 6,830,000 b) 0.000045

Exercise 6 You Try
Convert each of the following from standard notation to scientific notation.

a) 5,000,000,000 c) 0.001

b) 0.0000762 d) 438

Scientific Notation and Calculators

Calculators and computers display scientific notation in a slightly different way. Instead of displaying the "×10" portion of the scientific notation, most calculators display only the coefficient and the exponent of 10 is labeled with an "E".

4.4 Scientific Notation

It is important to get used to how a particular calculator displays a number in scientific notation in order to interpret the results of a calculation correctly.

Here are two wonderful sites that you may want to check out. Both sites take you on an adventure in powers of ten.

- https://www.youtube.com/watch?v=0fKBhvDjuy0
- http://htwins.net/scale2

World View Note. Archimedes (287 BC - 212 BC), the Greek mathematician, developed a system for representing large numbers using a system very similar to scientific notation. He used his system to calculate the number of gains of sand it would take to fill the universe. His conclusion was 10^{63} grains of sand because he figured the universe to have a diameter of 10^{14} stadia or about 2 light years.

4.4: Exercises

Write each number in scientific notation.

1. 8,850
2. 0.081
3. 0.00000391
4. 0.000744
5. 1,090,000
6. 15,000,000,000

Write each number in standard notation.

7. 8.7×10^5
8. 9×10^{-4}
9. 2×10^8
10. 2.56×10^2
11. 5.33×10^4
12. 6.7×10^{-5}

Complete the following table.

	Name	Standard Notation	Scientific Notation
13.	Trillion	1,000,000,000,000	
14.	Billion		1×10^9
15.	Million		1×10^6
16.	Thousand	1,000	
17.	Tenth	0.1	
18.	Hundredth	0.01	
19.	Millionth		1×10^{-6}
20.	Billionth		1×10^{-9}

Insert either < or > to make the statement true.

21. $2 \times 10^5 \;\square\; 2 \times 10^6$
22. $3 \times 10^5 \;\square\; 3 \times 10^{-5}$
23. $5 \times 10^{-5} \;\square\; 5 \times 10^{-6}$
24. $8 \times 10^5 \;\square\; 7 \times 10^5$
25. $2 \times 10^5 \;\square\; 2.1 \times 10^5$
26. $2 \times 10^6 \;\square\; 6 \times 10^2$

Perform the indicated operation. Write the answer in scientific notation.

27. $(4.7 \times 10^5)(2 \times 10^{-3})$
28. $(3.1 \times 10^{-6})^2$
29. $1.5(2.3 \times 10^{-5})$
30. $\dfrac{4.8 \times 10^8}{2 \times 10^5}$

Word Problems.

31. As of December 13, 2015, the national debt was nearly $19 trillion. Write this amount in scientific notation. (source: www.usdebtclock.org)

4.4 Scientific Notation

32. It is estimated that internet traffic will grow to 88 exabytes per month in 2016, or 88,000,000,000,000,000,000 bytes. Write this amount in scientific notation. (source: www.cisco.com)

33. Some hummingbirds beat their wings at a rate of 80 times/second. This is 0.0125 seconds/beat. Write this amount in scientific notation. (source: www.hummingbirds.net)

34. In 2015, it was estimated that Mark Zuckerberg's shares in Facebook were worth $\$4.5 \times 10^{10}$. Zuckerberg wants to give 99% of this Facebook shares away. If Zuckerberg keeps 1% of his Facebook shares, what is this value? Write the answer in both scientific and standard notation.

35. A nanometer is 1×10^{-9} meters. A kilometer is 1×10^3 meters. How many nanometers are in a kilometer?

36. For a typical 3 minute song, the mp3 file is 2.75×10^6 bytes. How many mp3 files could be stored in 32 gigabytes? Write the answer in both scientific and standard notation. (1 gigabyte $= 1 \times 10^9$ bytes)

 **Rescue Roody!**

37. Rewrite 0.00325 in scientific notation. Roody's answer: 5.23×10^{-4}.

38. Rewrite 1.23×10^{-5} in standard notation. Roody's answer: 123,000.

Chapter 4 Assessment

Simplify the following expressions.

1. 9^2

2. $(-9)^2$

3. -9^2

4. 9^{-2}

5. $(-9)^{-2}$

6. -9^{-2}

7. -3^0

8. $3x^0$

9. $\dfrac{3^0}{4}$

10. $\left(\dfrac{3}{4}\right)^0$

11. $3^{-2} - 4^0$

12. $\left(\dfrac{m^3}{7}\right)^2$

13. $(-2xy^2)(5xy)$

14. $\left(\dfrac{3}{w^2}\right)^{-4}$

15. $\dfrac{(n^{-3})^{-3}}{n}$

16. $8p^{-2}$

17. $\dfrac{6a^3}{4a^5}$

18. $\dfrac{-5n^3}{10n^2}$

Write the following in standard notation.

19. 2.84×10^8

20. 7.152×10^{-5}

Write the following in scientific notation.

21. $7,340,000,000,000$

22. 0.0000561

Perform the indicated operation. Write your answer in scientific notation.

23. $\dfrac{3.6 \times 10^4}{1.2 \times 10^{-2}}$

5. Polynomials

5.1 Introduction to Polynomials

Objective: To introduce polynomials

What is a polynomial?

The word **polynomial** is made up of two root words, poly meaning *many* and nomial meaning *terms*. A **term** is a real number or a real number times a variable to a positive integer power. A **polynomial** is a sum of one or more terms. Note that this implies that no variable can be part of a denominator.

> **Example 1** Determine whether the following is a polynomial or not. Explain why or why not.
>
> a) $4x^2 - 5$
>
> **Solution.**
> This is a polynomial because it is composed of a real number and a variable such that the variable has a positive integer exponent.
>
> b) $y^4 + y^{-2} + 8$
>
> **Solution.**
> This is not a polynomial because one of the variable's exponent is not a positive integer.

Exercise 1 Class Example
Determine whether the following is a polynomial or not.

a) $-2.5c^2 - c + 6$

c) $7g^{\frac{1}{2}}$

b) $\dfrac{5}{m+3}$

d) $\dfrac{p}{3} + \dfrac{1}{4}$

Exercise 2 You Try
Determine whether the following is a polynomial or not.

a) $4 + x^{1/3}$

c) $5 - \dfrac{6}{h}$

b) $t^3 - \dfrac{5}{8}$

d) $g^2 + 4g - 2.3$

Terms of a Polynomial

Terms can be connected by addition or subtraction. Expressions are often defined by the number of terms in the expression. The following are the most common classification of polynomials.

- **Monomial** - a single term such as $-3x^5$, p^3, 1.8, $\dfrac{1}{2}x$

- **Binomial** - two unlike terms held together by addition or subtraction such as $m^2 - m$, $0.3a + 4$, $6n^4 + \dfrac{5}{8}$

- **Trinomial** - three unlike terms held together by addition or subtraction such as $x^2 + x + 8$, $\dfrac{7}{3} - n + 2.3n^5$, $a + b + c$

- Expressions with more than three terms are simply called **polynomials**.

5.1 Introduction to Polynomials

Degree of a Polynomial

The **degree of a polynomial in one variable** is the highest exponent of that variable. For example, the degree of the polynomial, $6x^2 + 5x - 3$ is 2 because the highest exponent of the variable x is 2.

Example 2 Complete the table by writing the number of terms of the polynomial, degree of the polynomial and classifying the type of polynomial as monomial, binomial, etc.

Expression	Number of Terms	Degree of Polynomial	Type of Polynomial
$z^2 - 5z + 8$	3	2	Trinomial
7.9	1	0	Monomial
$\frac{5}{7} - h^3$	2	3	Binomial

Exercise 3 Class Example
Complete the table by writing the number of terms of the polynomial, degree of the polynomial and classifying the type of polynomial as monomial, binomial, etc.

Expression	Number of Terms	Degree of Polynomial	Type of Polynomial
$2 - 4c$			
$p^2 - \frac{1}{5}$			
$8m^4 + m^3 - 5m^2 - m + 1$			
$k + k^5 - \sqrt{3}$			
24			

Exercise 4 You Try

Complete the table by writing the number of terms of the polynomial, degree of the polynomial and classifying the type of polynomial as monomial, binomial, etc.

Expression	Number of Terms	Degree of Polynomial	Type of Polynomial
$2m^4$			
1			
$7 - 5p + p^2$			
$\frac{1}{8} - g^3$			
$m^3 + m^2 - 6m + 1$			

5.1: Exercises

Determine whether each of the following expressions is a polynomial or not.

1. $c^5 - 3c^4 + 8c - 1$
2. $b^3 - 9 + \dfrac{1}{b}$
3. $\dfrac{1}{2}w^2 + w - \dfrac{4}{5}$
4. $\sqrt{25x^2 + 9}$
5. π
6. $\dfrac{2+n}{n+7}$

Complete the table by writing the number of terms of the polynomial, the degree of the polynomial, and classifying the polynomial as a monomial, binomial, etc.

	Expression	Number of Terms	Degree	Type of Polynomial
7.	$x^2 + 9.1x + 2.3$			
8.	$3n - 4n^4$			
9.	$-\dfrac{5}{8}$			
10.	$n^3 - 7n^2 + 15n - 20$			
11.	$2p$			
12.	$n^7 + n^3$			

5.2 Addition and Subtraction of Polynomials

Objective: To add, subtract and evaluate polynomials

Like Terms

Polynomials can be added and subtracted if there are like terms in the expression.
What are like terms? They are terms whose variable and its exponent are the same. Lets look at some examples:

- $5x$ and $-2x$ are like terms since both terms have the same variable x with the same exponent, 1.

- $5x$ and $2y$ are unlike terms because the variables in each term are different.

- The terms $5x$ and $5x^2$ are also unlike terms because the exponents of the variable x in each term are not the same even though both terms have the same variable, x.

Exercise 1 Class Example
Determine if the following terms are like or unlike. Explain your reasoning.

Terms	Like or Unlike	Explanation
$3y$ and y		
$4x^2$ and $-6x^2$		
$7w$ and $5w^3$		
$9a$ and $4b$		
2 and z		

To add (or subtract) like terms, we add (or subtract) the coefficients and keep the same variable and its exponent.

5.2 Addition and Subtraction of Polynomials

Example 1 Complete the table by finding the sum or difference of each binomial.

Given	Sum or Difference	Explanation
$2x + 6x$	$= 8x$	Like terms; add coefficients
$4x - 6x^2$	$= 4x - 6x^2$	Unlike terms; exponents of variable are different
$5y + 3z$	$= 5y + 3z$	Unlike terms; variables are different
$7x^3 - 8x^3$	$= -x^3$	Like terms; subtract coefficients

Exercise 2 Class Example
Complete the table by finding the sum or difference of each binomial and explaining what you did.

Given	Sum or Difference	Explanation
$9x^2 + x$		
$5x^2 - 4x^2$		
$2a + 8b$		
$\dfrac{3}{4}w - \dfrac{5}{6}w$		
$x^3 + \dfrac{2}{7}x^2$		

Exercise 3 You Try
Complete the table by finding the sum or difference of each binomial and explaining what you did.

Given	Sum or Difference	Explanation
$2x^2 - x^2$		
$x^2 - 3.5x$		
$7x - 1.8$		
$5y + 3z$		
$\frac{2}{3}x + \frac{1}{6}x$		

Adding and Subtracting Polynomials

If we have two (or more) polynomials we can add (or subtract) them. Remember that we can only add and subtract like terms. We **cannot solve** for the variable but we can combine like terms. Generally, the answer is written in descending order with the highest power of the variable written first, then in order from greatest to least.

Example 2 Perform the indicated operation given $(4x^2 - 2x + 8) + (3x^2 - 9x - 11)$. Write the answer in descending order.

Solution.

$(4x^2 - 2x + 8) + (3x^2 - 9x - 11)$ Remove parenthesis first
$= 4x^2 - 2x + 8 + 3x^2 - 9x - 11$ Combine like terms
$= 7x^2 - 11x - 3$ Our Solution

Exercise 4 Class Example
Perform the indicated operation given $(7x^2 + 2x + 5) + (6x^2 - 8x - 9)$. Write your answer in descending order.

5.2 Addition and Subtraction of Polynomials

Exercise 5 **You Try**
Perform the indicated operation given $(8w^2 + 6w) + (4w^2 - 6w + 5)$. Write your answer in descending order.

When we subtract polynomials, we must remember to distribute the subtraction sign to all terms within the parentheses. Remember that a subtraction sign before a parenthesis is the same as having -1 before the parenthesis. Subtraction will change all the signs within the parenthesis.

Example 3 Perform the indicated operation given $(5x^2 - 2x + 7) - (3x^2 + 6x - 4)$. Write the answer in descending order.

Solution.

$(5x^2 - 2x + 7) - (3x^2 + 6x - 4)$ Distribute -1 through second parenthesis
$\quad = 5x^2 - 2x + 7 - 3x^2 - 6x + 4$ Combine like terms
$\quad = 2x^2 - 8x + 11$ Our Solution

Exercise 6 **Class Example**
Perform the indicated operation given $(2x^3 - 3x^2 - 6x + 5) - (3x^3 - 9x^2 - 8x + 7)$. Write your answer in descending order.

Exercise 7 **You Try**
Perform the indicated operation and write your answer in descending order.

a) $(7m^2 - 5m - 1) - (m^2 + 8m - 1)$

b) $(2y^3 - 5y^2 + 7y + 5) - (-y^3 + y^2 - 8y + 1)$

Let us take a look at an example of addition and subtraction of polynomials when combined in the same problem.

Example 4 Perform the indicated operation given $(2x^2 - 4x + 3) + (5x^2 - 6x + 1) - (x^2 - x + 1)$. Write the answer in descending order.

Solution.

$(2x^2 - 4x + 3) + (5x^2 - 6x + 1) - (x^2 - x + 1)$ Distribute -1 through third parenthesis
$= 2x^2 - 4x + 3 + 5x^2 - 6x + 1 - x^2 + x - 1$ Combine like terms
$= 6x^2 - 9x + 3$ Our Solution

Exercise 8 Class Example
Perform the indicated operation given $(5m^2 + 2m - 7) - (m^2 - m + 1) + (3m^2 + m - 2)$. Write your answer in descending order.

Exercise 9 You Try
Perform the indicated operation given $(y^2 - 3y - 4) + (5y^2 + 6) - (7y^2 - 8y - 1)$. Write your answer in descending order.

5.2: Exercises

Combine like terms. Write answers in descending order.

1. $x^2 + 3x^2$
2. $2x^2 - x$
3. $6x^3 + 4x^2$
4. $5k^2 + 3k - 2k^2 + 6k$
5. $(x^2 + 5x^3) + (7x^2 + 3x^3)$
6. $(6x^3 + 5x) - (8x + 6x^3)$
7. $(4n^4 + 2n^2 + 3) + (2n^4 - 7n^2 - 4)$
8. $(4p^2 - 2p - 3) - (3p^2 - 6p + 3)$
9. $(4b^3 + 7b^2 - 3b) + (b^3 + 5b^2 + 8)$
10. $(7 + 4m + 8m^2) - (5m^2 + 6m + 1)$
11. $(4x^3 - 7x^2 + x) + (6x^3 + 7x^2 + 2x - 8)$
12. $(2 - 2n^2 + 7n^4) + (2 + 2n^2 + 4n^3 + 2n^4)$
13. $(7b^3 - b + 8) - (3b^3 + 7b^2 + 7b - 8) + (6b^2 - 3b + 3)$
14. $(3n^3 - 8n^2 - 1) + (7n^3 + 3n^2 - 6n + 2) + (4n^3 + 8n^2 + 7)$
15. $(8x^3 + 2x^2 - 2x + 9) - (3x^3 + 2x^2 - x + 1) - (5x^3 - x + 8)$
16. $(4x^3 + 7x^2 - 2x + 8) - (2x^3 - 6x^2 + 8) + (5x^3 - 4x^2 + 6x)$

Rescue Roody!
Roody is learning how to combine like terms. Unfortunately, none of his work is correct. Help Roody.

17. Combine like terms: $x^2 - x$. Roody's work:

    $x^2 - x = x$

18. Combine like terms: $(4x - 5) - (2x + 7)$. Roody's work:

    $(4x - 5) - (2x + 7) = 4x - 5 - 2x + 7$
    $= 2x + 2$

5.3 Multiplication of Polynomials

Objective: To multiply polynomials

Multiplication by Monomials

To multiply two monomials, we multiply coefficients and then the variables, following the product rule for exponents. Lets look at the difference between adding or subtracting monomials and multiplying two monomials.

Example 1 Perform the indicated operation.

a) $4x^3 + 3x^2$

b) $(4x^3)(3x^2)$

Solution.

a) $4x^3 + 3x^2$ - Unlike terms; cannot combine terms

b)
$$(4x^3)(3x^2) = (4 \cdot x \cdot x \cdot x) \cdot (3 \cdot x \cdot x) \quad \text{Expand}$$
$$= (4 \cdot 3) \cdot (x \cdot x \cdot x \cdot x \cdot x \cdot) \quad \text{Use the Commutative Property}$$
$$= 12x^5 \quad \text{Our Solution}$$

Exercise 1 Class Example
Perform the indicated operation.

a) $5x^3 + x^3$

b) $(5x^3)(x^3)$

c) $x^2 - x^3$

d) $(x^2)(-x^3)$

5.3 Multiplication of Polynomials

Exercise 2 You Try
Perform the indicated operation.

a) $x + 4x^2$

b) $(x)(4x^2)$

c) $6x^2 - 7x^2$

d) $(6x^2)(-7x^2)$

Next, we will consider multiplying a polynomial by a monomial. To find the product, we will use the distributive property, that is, we will multiply each term of the polynomial by the monomial.

Example 2 Multiply $4x^3(5x^2 - 2x + 5)$

Solution.

$$4x^3(5x^2 - 2x + 3) = (4x^3)(5x^2) - (4x^3)(2x) + (4x^3)(3) \qquad \text{Distribute } 4x^3$$
$$= 20x^5 - 8x^4 + 12x^3 \qquad \text{Our Solution}$$

Exercise 3 Class Example
Multiply the following.

a) $4x(3x + 7)$

b) $-y(8y^2 - 9)$

c) $(6n - 7) \cdot 3n^2$

d) $5p^2(6p^2 - 2p + 5)$

Exercise 4 You Try
Perform the indicated operation.

a) $-3(4r-5)$

b) $(n^2+3) \cdot 2n^2$

c) $8w(w^2+4w-6)$

d) $-3y^2(6y^3-2y+7)$

Multiplication of Binomials

We will now consider multiplying two binomials. This is done by multiplying each term in the first binomial by each term in the second binomial.

Example 3 Multiply $(4x+7)(3x-2)$

Solution.

$$(4x+7)(3x-2) = 4x(3x-2)+7(3x-2) \qquad \text{Distribute}$$
$$= 12x^2-8x+21x-14 \qquad \text{Combine like terms}$$
$$= 12x^2+13x-14 \qquad \text{Our Solution}$$

In short, when multiplying two binomials we need to multiply each term of the first binomial by each term of the second binomial, giving a total of 4 terms. After this multiplication is done we should combine any like terms.

In many textbooks, multiplying two binomials is known as the FOIL method. The letters of FOIL help us remember every combination of the binomial multiplication.

- **F** stands for First - we multiply the first term of each binomial.

- **O** stands for Outside - we multiply the outside two terms.

- **I** stands for Inside - we multiply the inside two terms.

- **L** stands for Last - we multiply the last term of each binomial.

5.3 Multiplication of Polynomials

Example 4 Multiply $(4x+7)(3x-2)$ using the FOIL method

Solution.
Using the FOIL method to multiply, we get the following.

$(4x)(3x) = 12x^2$ F - multiply First terms

$(4x)(-2) = -8x$ O - multiply Outside terms

$(7)(3x) = 21x$ I - multiply the Inside terms

$(7)(-2) = -14$ L - multiply the Last terms

Therefore,

$(4x+7)(3x-2) = 12x^2 - 8x + 21x - 14$ Combine like terms
$= 12x^2 + 13x - 14$ Our Solution

Exercise 5 Class Example
Perform the indicated operation.

a) $(x+1)(x-4)$

b) $(2y+3)(y+2)$

c) $(7n+6)(7n-6)$

d) $(2a+3b)(8a-7b)$

Exercise 6 You Try
Perform the indicated operation.

a) $(x+3)(x+8)$

b) $(n-9)(5n+7)$

c) $(3m+2)(3m-2)$

d) $(4x+3y)(5x+8y)$

A polynomial base raised to a positive integer exponent works exactly the same way as a variable base raised to a positive integer exponent. That is, the exponent tells us how many times the base is multiplied by itself, whether the base is a variable or a polynomial.

Example 5 Multiply $(2x+5)^2$

Solution.
The base in this case is a binomial, $2x+5$, and it is raised to the power 2. This means we have to multiply the base by itself, or $(2x+5)(2x+5)$.

$(2x+5)^2 = (2x+5)(2x+5)$ Expand $(2x+5)^2$

$ = (2x)(2x)+(2x)(5)+(5)(2x)+(5)(5)$ FOIL

$ = 4x^2+10x+10x+25$ Combine like terms

$ = 4x^2+20x+25$ Our Solution

Warning. $(2x+5)^2 \neq 4x^2+25$ as shown in the example above. The power rule for exponent does not apply when terms are added or subtracted.

5.3 Multiplication of Polynomials

Example 6 Multiply $(y-2)^3$

Solution.
The base in this case is a binomial, $y-2$, and it is raised to the power 3. This means we have to multiply the base by itself three times, or $(y-2)(y-2)(y-2)$.

$$
\begin{aligned}
(y-2)^3 &= (y-2)(y-2)(y-2) & &\text{Expand } (y-2)^3 \\
&= (y^2 - 2y - 2y + 4)(y-2) & &\text{Multiply first 2 factors} \\
&= (y^2 - 4y + 4)(y-2) & &\text{Combine like terms} \\
&= y^2(y-2) - 4y(y-2) + 4(y-2) & &\text{Distribute} \\
&= y^2(y) + y^2(-2) - 4y(y) - 4y(-2) + 4(y) + 4(-2) & &\text{Multiply} \\
&= y^3 - 2y^2 - 4y^2 + 8y + 4y - 8 & &\text{Combine like terms} \\
&= y^3 - 6y^2 + 12y - 8 & &\text{Our Solution}
\end{aligned}
$$

Warning. $(y-2^3) \neq y^3 - 8$ as shown in the example above. The power rule for exponents does not apply when terms are added or subtracted.

Exercise 7 Class Example
Multiply the following.

a) $(a-4)^2$

b) $(2m-3)(4m^2 + 4m + 5)$

c) $(y+1)^3$

d) $3(x+6)(2x-5)$

Exercise 8 You Try
Multiply the following.

a) $(n+6)^2$

b) $-2(x-5)^2$

c) $(6a+7)(2a^2+5a-1)$

d) $(x-4)^3$

Example 7 Perform the indicated operation on $(x+5)(x-5)-(x-3)^2$ and write the answer in descending order.

Solution.
Remember to follow the order of operation.

$(x+5)(x-5)-(x-3)^2$	Expand $(x-3)^2$
$=(x+5)(x-5)-(x-3)(x-3)$	Multiply
$=(x^2-5x+5x-25)-(x^2-3x-3x+9)$	Combine like terms
$=(x^2-25)-(x^2-6x+9)$	Distribute -1 through second parenthesis
$=x^2-25-x^2+6x-9$	Combine like terms
$=6x-34$	Our Solution

5.3 Multiplication of Polynomials

Exercise 9 Class Example

Perform the indicated operation on $(2x+1)^2 - (x-4)(x+4)$ and write the answer in descending order.

Exercise 10 You Try

Perform the indicated operation on $2(x+1)^2 - (2x-3)^2$ and write the answer in descending order.

5.3: Exercises

Perform the indicated operation.

1. $6(p-7)$
2. $2x(6x+3)$
3. $5m^4(4m+4)$
4. $(c+3)(c-5)$
5. $(x+5)(x+3)$
6. $(3v-4)(5v-2)$
7. $(6x-7)(4x+1)$
8. $(x+3y)(3x+4y)$
9. $(a+b)(a-b)$
10. $(3-y)(3+y)$
11. $(5n-4)(5n+4)$
12. $(3p-7)^2$
13. $(a+b)^2$
14. $(w+2)^3$
15. $(c-1)^3$
16. $(a-b)^3$
17. $(r-7)(6r^2-r+5)$
18. $7(x-5)(x-2)$
19. $6(4x-1)(4x+1)$
20. $2(3n-2)(2n^2-2n+5)$
21. $3x(x-4)+(x-3)^2$
22. $(6-y)^2-2y(5y+4)$
23. $(x+5)^2-(x+6)(x-2)$
24. $(2x-1)^2-(x+1)^2$

Rescue Roody!
Roody was told to perform the indicated operation, but he keeps getting wrong answers. Here is Roody's work. Help Roody.

25. $(x-9)^2 = x^2-81$

26. $(x+8)^2 = (x+8)(x+8) = x^2+64$

5.4 Division of Polynomials

Objective: To divide polynomials

Dividing by a Monomial

To divide a polynomial by a monomial, we divide each term of the polynomial by the monomial and simplify the resulting expression using the quotient rule for exponents.

For example, if we were dividing integers, we know that $\frac{12}{2}$ can be simplified as 6. If $\frac{12}{2}$ was written as $\frac{2+4+6}{2}$ instead, to simplify, we divide each term in the numerator by the denominator, 2. That is,

$$\frac{2}{2} + \frac{4}{2} + \frac{6}{2} = 1 + 2 + 3 = 6$$

It is important to remember that each term in the numerator has to be divided by the denominator.

Example 1 Simplify $\dfrac{6x^4 - 18x^3 + 3x^2}{3x^2}$

Solution.

$\dfrac{6x^4 - 18x^3 + 3x^2}{3x^2}$ Divide each term in the numerator by the denominator, $3x^2$

$= \dfrac{6x^4}{3x^2} - \dfrac{18x^3}{3x^2} + \dfrac{3x^2}{3x^2}$ Apply quotient rule for exponents

$= 2x^2 - 6x + 1$ Our Solution

Verify that we have the correct solution by multiplying the quotient by the divisor to see if our answer matches the dividend.

$3x^2(2x^2 - 6x + 1) = 6x^4 - 18x^3 + 3x^2$ ✓

Note. Whenever we divide a polynmial by a monomial, our answer will be a polynomial with the same number of terms as the original polynomial. In the above example, the polynomial in the numerator has 3 terms. Therefore, the solution should also contain 3 terms. A common mistake is to ignore the last term because the monomial goes evenly into the last term. Remember that $\dfrac{3x^2}{3x^2} = 1$, not 0.

Exercise 1 Class Example

Simplify $\dfrac{8x^3 + 4x^2 - 2x + 6}{4x^2}$

Exercise 2 You Try

Simplify each of the following.

a) $\dfrac{25k^3 + 15k^2 - 5k}{5k}$

b) $\dfrac{16m^3 + 12m^2 - 4m + 24}{12m^2}$

Long Division

Lets first review long division with whole numbers. The same process will be used to do long division with polynomials.

Example 2 Divide $631 \div 4$ using long division.

Solution.
Rewrite $631 \div 4$ in long division form, $4\overline{)631}$.

Divide the first digit of the quotient by the divisor, if possible. We see that $\frac{6}{4} = 1$ with a remainder of 2. Multiply the quotient (excluding the remainder) by the divisor to get $1 \cdot 4 = 4$. Subtract this result from the first digit of the dividend and then bring down the next digit. This is our new dividend.

$$\begin{array}{r} 1 \\ 4\overline{)631} \\ \underline{4} \\ 23 \end{array}$$

Divide the resulting difference by the divisor. We get $\frac{23}{4} = 5$ with a remainder of 3. Multiply the quotient (excluding the remainder) by the divisor to get $5 \cdot 4 = 20$. Subtract this result from the

5.4 Division of Polynomials

new dividend and then bring down the next digit. This is our new dividend.

$$
\begin{array}{r}
15 \\
4\overline{)631} \\
\underline{4} \\
23 \\
\underline{20} \\
3\,1
\end{array}
$$

Divide the resulting difference by the divisor. We get $\frac{31}{4} = 7$ with a remainder of 3. Multiply the quotient (excluding the remainder) by the divisor to get $7 \cdot 4 = 28$. Subtract this result from the new dividend and then bring down the next digit, if any.

$$
\begin{array}{r}
157 \\
4\overline{)631} \\
\underline{4} \\
23 \\
\underline{20} \\
31 \\
\underline{28} \\
3
\end{array}
$$

We are now done with long division because are no more digits to bring down and the difference is smaller than the divisor. Therefore, $631 \div 4 = 157 + \frac{3}{4}$

Verify that we have the correct solution.

$157 \cdot 4 = 628$ Multiply quotient (exclude remainder) by divisor

$682 + 3 = 631$ ✓   Add remainder part; answer matches dividend

Long division is required when we divide by polynomials which are not monomials. Long division with polynomials is very similar to long division with whole numbers.

Steps in Dividing Polynomials Using Long Division

Be sure to write the polynomials in descending order before dividing. Replace missing terms with a 0 or skip a space.

Step 1. Divide the first term of the dividend by the first term of the divisor.

Step 2. Multiply the quotient by the divisor.

Step 3. Subtract the result from the dividend.

Step 4. Bring down the next term. The difference and the next term will be the new dividend.

Step 5. Repeat the whole process until the degree of the difference is smaller than the divisor.

Example 3 Divide $\dfrac{x^2+2x-12}{x+5}$ using long division.

Solution.

The divisor, $x+5$, is not a monomial. We will do polynomial long division. First, rewrite $\dfrac{x^2+2x-12}{x+5}$ in long division form. Be sure the terms are written in descending order.

$$x+5 \overline{\smash{\big)}\, x^2+2x-12}$$

We will now go through the steps in polynomial long division.

Step 1. Divide the first term of the dividend by the first term of the divisor, $\dfrac{x^2}{x}=x$.

$$\begin{array}{r} x \\ x+5 \overline{\smash{\big)}\, x^2+2x-12} \end{array}$$

Step 2. Multiply the quotient, x, by the divisor to get $x(x+5)=x^2+5x$.

Step 3. Subtract (x^2+5x), that is, $-(x^2+5x)=-x^2-5x$. Write this result underneath the dividend. Be sure to line the terms up correctly.

$$\begin{array}{r} x \\ x+5 \overline{\smash{\big)}\, x^2+2x-12} \\ -x^2-5x \end{array}$$

Step 4. Perform the indicated operation and bring down the next term.

$$\begin{array}{r} x \\ x+5 \overline{\smash{\big)}\, x^2+2x-12} \\ -x^2-5x \\ \hline -3x-12 \end{array}$$

5.4 Division of Polynomials

Step 5. Since the new dividend's degree is equal to the divisor, we will repeat the whole process by going back to Step 1.

Step 1. Divide the first term of the new dividend with the first term of the divisor, $\dfrac{-3x}{x} = -3$.

$$\begin{array}{r} x\ -3 \\ x+5 \overline{\smash{)}\ x^2+2x-12} \\ \underline{-x^2-5x} \\ -3x-12 \end{array}$$

Step 2. Multiply the quotient, -3, by the divisor to get $-3(x+5) = -3x - 15$.

Step 3. Subtract $(-3x - 15)$, that is, $-(-3x - 15) = 3x + 15$. Write this result underneath the new dividend. Be sure to line the terms up correctly.

$$\begin{array}{r} x\ -3 \\ x+5 \overline{\smash{)}\ x^2+2x-12} \\ \underline{-x^2-5x} \\ -3x-12 \\ \underline{3x+15} \end{array}$$

Step 4. Perform the indicated operation and bring down the next term.

$$\begin{array}{r} x\ -3 \\ x+5 \overline{\smash{)}\ x^2+2x-12} \\ \underline{-x^2-5x} \\ -3x-12 \\ \underline{3x+15} \\ 3 \end{array}$$

Step 5. The degree of the result is now smaller than the divisor. We have a remainder, 3.

Therefore, $\dfrac{x^2+2x-12}{x+5} = x - 3 + \dfrac{3}{x+5}$

Verify that we have the correct solution.

$(x-3)(x+5)$	Multiply polynomial part of quotient by divisor
$= x^2 + 5x - 3x - 15$	Combine like terms
$= x^2 + 2x - 15$	
$= x^2 + 2x - 15 + 3$	Add remainder part
$= x^2 + 2x - 12 \quad \checkmark$	Answer matches dividend

Exercise 3 Class Example
Divide $\dfrac{x^2 - 3x - 53}{x - 9}$ using long division.

Exercise 4 You Try
Divide $\dfrac{x^2 - 10x + 16}{x - 3}$ using long division.

5.4 Division of Polynomials

Example 4 Divide $\dfrac{2x^3 - 5x^2 - 7x + 4}{2x + 1}$ using long division.

Solution.

The divisor, $2x + 1$, is not a monomial. We will do polynomial long division. First, rewrite $\dfrac{2x^3 - 5x^2 - 7x + 4}{2x + 1}$ in long division form. Be sure the terms are written in descending order.

$$2x + 1 \,\overline{\smash{)}\, 2x^3 - 5x^2 - 7x + 4}$$

We will now go through the steps in polynomial long division.

Step 1. Divide the first term of the dividend by the first term of the divisor, $\dfrac{2x^3}{2x} = x^2$.

$$\begin{array}{r} x^2 \\ 2x+1 \,\overline{\smash{)}\, 2x^3 - 5x^2 - 7x + 4} \end{array}$$

Step 2. Multiply the quotient, x^2, by the divisor to get $x^2(2x + 1) = 2x^3 + x^2$.

Step 3. Subtract $(2x^3 + x^2)$, that is, $-(2x^3 + x^2) = -2x^3 - x^2$. Write this result underneath the dividend. Be sure to line the terms up correctly.

$$\begin{array}{r} x^2 \\ 2x+1 \,\overline{\smash{)}\, 2x^3 - 5x^2 - 7x + 4} \\ \underline{-2x^3 - x^2 } \end{array}$$

Step 4. Perform the indicated operation and bring down the next term.

$$\begin{array}{r} x^2 \\ 2x+1 \,\overline{\smash{)}\, 2x^3 - 5x^2 - 7x + 4} \\ \underline{-2x^3 - x^2 } \\ -6x^2 - 7x \end{array}$$

Step 5. Since the new dividend's degree is greater than the divisor, we will repeat the whole process by going back to Step 1.

Step 1. Divide first term of the new dividend with the first term of the divisor, $\dfrac{-6x^2}{2x} = -3x$.

$$\begin{array}{r} x^2 - 3x \\ 2x+1 \,\overline{\smash{)}\, 2x^3 - 5x^2 - 7x + 4} \\ \underline{-2x^3 - x^2 } \\ -6x^2 - 7x \end{array}$$

Step 2. Multiply the quotient, $-3x$, by the divisor to get $-3x(2x+1) = -6x^2 - 3x$.

Step 3. Subtract $(-6x^2 - 3x)$, that is, $-(-6x^2 - 3x) = 6x^2 + 3x$. Write this result underneath the new dividend. Be sure to line the terms up correctly.

$$
\begin{array}{r}
x^2 - 3x \\
2x+1 \overline{)\ 2x^3 - 5x^2 - 7x + 4} \\
\underline{-2x^3 - x^2 } \\
-6x^2 - 7x \\
\underline{6x^2 + 3x }
\end{array}
$$

Step 4. Perform the indicated operation and bring down the next term.

$$
\begin{array}{r}
x^2 - 3x \\
2x+1 \overline{)\ 2x^3 - 5x^2 - 7x + 4} \\
\underline{-2x^3 - x^2 } \\
-6x^2 - 7x \\
\underline{6x^2 + 3x } \\
-4x + 4
\end{array}
$$

Step 5. Since the new dividend's degree is equal to the divisor, we will repeat the whole process by going back to Step 1.

Step 1. Divide the first term of the new dividend with the first term of the divisor, $\dfrac{-4x}{2x} = -2$.

$$
\begin{array}{r}
x^2 - 3x - 2 \\
2x+1 \overline{)\ 2x^3 - 5x^2 - 7x + 4} \\
\underline{-2x^3 - x^2 } \\
-6x^2 - 7x \\
\underline{6x^2 + 3x } \\
-4x + 4
\end{array}
$$

Step 2. Multiply the quotient, -2, by the divisor to get $-2(2x+1) = -4x - 2$.

Step 3. Subtract $(-4x - 2)$, that is, $-(-4x - 2) = 4x + 2$. Write this result underneath the dividend. Be sure to line the terms up correctly.

$$
\begin{array}{r}
x^2 - 3x - 2 \\
2x+1 \overline{)\ 2x^3 - 5x^2 - 7x + 4} \\
\underline{-2x^3 - x^2 } \\
-6x^2 - 7x \\
\underline{6x^2 + 3x } \\
-4x + 4 \\
\underline{4x + 2}
\end{array}
$$

5.4 Division of Polynomials

Step 4. Perform the indicated operation and bring down the next term.

$$\begin{array}{r} x^2-3x-2 \\ 2x+1{\overline{\smash{\big)}\,2x^3-5x^2-7x+4}} \\ \underline{-2x^3-x^2} \\ -6x^2-7x \\ \underline{6x^2+3x} \\ -4x+4 \\ \underline{4x+2} \\ 6 \end{array}$$

Step 5. The degree of the result is now smaller than the divisor. We have a remainder, 6.

Therefore, $\dfrac{2x^3-5x^2-7x+4}{2x+1} = x^2-3x-2+\dfrac{6}{2x+1}$

Verify that we have the correct solution.

$(x^2-3x-2)(2x+1)$ Multiply polynomial part of quotient by divisor
$= 2x^3+x^2-6x^2-3x-4x-2$ Combine like terms
$= 2x^3-5x^2-7x-2$
$2x^3-5x^2-7x-2+6$ Add remainder part
$= 2x^3-5x^2-7x+4$ ✓   Answer matches dividend

Exercise 5   Class Example

Divide $\dfrac{6x^3-5x^2-8x-2}{3x-1}$ using long division.

Exercise 6 You Try
Divide the following using long division.

a) $\dfrac{x^3 - x^2 - 16x + 8}{x - 4}$

b) $\dfrac{6x^3 + 7x^2 + 11x - 12}{2x + 3}$

Example 5 Divide $\dfrac{2x^3 + 42 - 4x}{x + 3}$ using long division.

Solution.

The divisor, $x + 3$, is not a monomial. We will do polynomial long division. Note that the dividend is not written in descending order. It is also missing the x^2-term. We can either put a 0 or skip a space where the x^2-term should be. Let us rewrite the problem in long division form.

$$x+3 \overline{)\ 2x^3 \qquad\quad -4x + 42}$$

We will now go through the steps in polynomial long division.

Step 1. Divide the first term of the dividend by the first term of the divisor, $\dfrac{2x^3}{x} = 2x^2$.

$$\begin{array}{r} 2x^2 \qquad\qquad\quad \\ x+3 \overline{)\ 2x^3 \qquad\quad -4x + 42} \end{array}$$

Step 2. Multiply the quotient, $2x^2$, by the divisor to get $2x^2(x+3) = 2x^3 + 6x^2$.

Step 3. Subtract $(2x^3 + 6x^2)$, that is, $-(2x^3 + 6x^2) = -2x^3 - 6x^2$. Write this result underneath the dividend. Be sure to line the terms up correctly.

5.4 Division of Polynomials

$$\begin{array}{r} 2x^2 \\ x+3 \overline{\smash{\big)}\, 2x^3 -4x+42} \\ \underline{-2x^3-6x^2 } \end{array}$$

Step 4. Perform the indicated operation and bring down the next term.

$$\begin{array}{r} 2x^2 \\ x+3 \overline{\smash{\big)}\, 2x^3 -4x+42} \\ \underline{-2x^3-6x^2 } \\ -6x^2 \; -4x \end{array}$$

Step 5. Since the new dividend's degree is greater than the divisor, we will repeat the whole process by going back to Step 1.

Step 1. Divide the first term of the new dividend with the first term of the divisor, $\dfrac{-6x^2}{x} = -6x$.

$$\begin{array}{r} 2x^2 \; -6x \\ x+3 \overline{\smash{\big)}\, 2x^3 -4x+42} \\ \underline{-2x^3-6x^2 } \\ -6x^2 \; -4x \end{array}$$

Step 2. Multiply the quotient, $-6x$, by the divisor to get $-6x(x+3) = -6x^2 - 18x$.

Step 3. Subtract $(-6x^2 - 18x)$, that is, $-(-6x^2 - 18x) = 6x^2 + 18x$. Write this result underneath the new dividend. Be sure to line the terms up correctly.

$$\begin{array}{r} 2x^2 \; -6x \\ x+3 \overline{\smash{\big)}\, 2x^3 -4x+42} \\ \underline{-2x^3-6x^2 } \\ -6x^2 \; -4x \\ 6x^2 + 18x \end{array}$$

Step 4. Perform the indicated operation and bring down the next term.

$$\begin{array}{r} 2x^2 \; -6x \\ x+3 \overline{\smash{\big)}\, 2x^3 -4x+42} \\ \underline{-2x^3-6x^2 } \\ -6x^2 \; -4x \\ \underline{6x^2 + 18x } \\ 14x + 42 \end{array}$$

Step 5. Since the new dividend's degree is equal to the divisor, we will repeat the whole process by going back to Step 1.

Step 1. Divide the first term of the new dividend with the first term of the divisor, $\frac{14x}{x} = 14$

$$
\begin{array}{r}
2x^2 - 6x + 14 \\
x+3 \overline{\smash{)}\, 2x^3 - 4x + 42} \\
\underline{-2x^3 - 6x^2 } \\
-6x^2 - 4x \\
\underline{6x^2 + 18x } \\
14x + 42
\end{array}
$$

Step 2. Multiply the quotient, 14, by the divisor to get $14(x+3) = 14x + 42$.

Step 3. Subtract $(14x + 42)$, that is, $-(14x + 42) = -14x - 42$. Write this result underneath the dividend. Be sure to line the terms up correctly.

$$
\begin{array}{r}
2x^2 - 6x + 14 \\
x+3 \overline{\smash{)}\, 2x^3 - 4x + 42} \\
\underline{-2x^3 - 6x^2 } \\
-6x^2 - 4x \\
\underline{6x^2 + 18x } \\
14x + 42 \\
-14x - 42
\end{array}
$$

Step 4. Perform the indicated operation and bring down the next term.

$$
\begin{array}{r}
2x^2 - 6x + 14 \\
x+3 \overline{\smash{)}\, 2x^3 - 4x + 42} \\
\underline{-2x^3 - 6x^2 } \\
-6x^2 - 4x \\
\underline{6x^2 + 18x } \\
14x + 42 \\
\underline{-14x - 42} \\
0
\end{array}
$$

Step 5. The degree of the result is now smaller than the divisor. We have no remainder.

Therefore, $\dfrac{2x^3 - 4x + 42}{x + 3} = 2x^2 - 6x + 14$

5.4 Division of Polynomials

Verify that we have the correct solution.

$(2x^2 - 6x + 14)(x + 3)$ Multiply the quotient by the divisor
$= 2x^3 + 6x^2 - 6x^2 - 18x + 14x + 42$ Combine like terms
$= 2x^3 - 4x + 42$ ✓   Answer matches dividend

Exercise 7 Class Example
Divide $\dfrac{x^3 + 22 - 46x}{x + 7}$ using long division.

Exercise 8 Class Example
Divide $\dfrac{2x^3 + 11x^2 - 19}{2x + 3}$ using long division.

5.4: Exercises

Perform the indicated operation.

1. $\dfrac{9m^4 + 18m^3 + 27m^2}{9m^2}$

2. $\dfrac{9x^4 + 24x^3 - 6x^2 - 3x}{3x}$

3. $\dfrac{20x^4 + x^3 + 2x^2}{4x^3}$

4. $\dfrac{20n^4 + 50n^3 + 10n^2 + 40n}{10n^2}$

5. $\dfrac{3k^3 + 4k^2 + 2k + 8}{8k}$

6. $\dfrac{5x^4 + 45x^3 + 4x^2 + 9x}{9x}$

7. $\dfrac{a^2 - 4a - 30}{a - 8}$

8. $\dfrac{v^2 - 2v - 80}{v - 10}$

9. $\dfrac{n^2 + 7n + 5}{n + 4}$

10. $\dfrac{6x^2 + 5x - 6}{3x - 2}$

11. $\dfrac{4r^2 + 8r + 5}{2r + 5}$

12. $\dfrac{x^3 - 16x^2 + 71x - 56}{x - 8}$

13. $\dfrac{k^3 - 4k^2 - 6k + 4}{k - 1}$

14. $\dfrac{n^2 - 4}{n - 2}$

15. $\dfrac{9v^2 + 2}{3v - 1}$

16. $\dfrac{a^3 - 15a - 22}{a + 2}$

17. $\dfrac{4m^3 - 13m^2 - 9}{m - 3}$

18. $\dfrac{2n^3 + 11n^2 - 18}{2n + 3}$

Chapter 5 Assessment

For each problem below, do the following.

- Perform the indicated operation.
- Write your answer in descending order.
- Identify the degree of the resulting polynomial.
- State whether the resulting polynomial is a monomial, binomial, trinomial, or none of the above.

1. $(4x^2 + 3x - 7) + (x^2 - 3x + 9)$

2. $(7n^3 + 2n^2 - 9n + 8) - (3n^3 - 2n^2 - 9n + 4)$

3. $(2y^2 - 7y + 8) - (6y^2 + 6y - 8) + (4y^2 - 2y + 3)$

4. $2n^2(3n - 8)$

5. $(m + 8)(m - 7)$

6. $(2w - 5)(3w - 4)$

7. $\left(h + \dfrac{3}{4}\right)\left(h - \dfrac{3}{4}\right)$

8. $(p - 5)^2$

9. $(y + 4)(y^2 + 2y - 5)$

10. $3(2x + 1)(x^2 + 6x - 1)$

Divide each of the following expressions.

11. $\dfrac{6m^2 + 12m - 9}{3m}$

12. $\dfrac{15x^3 - 10x^2 + 5x}{10x^2}$

13. $\dfrac{3x^2 + x - 10}{x + 2}$

14. $\dfrac{2x^3 - 9x^2 - 17x + 39}{2x - 3}$

Answer Key

1.1: Answers

1. 24
2. $\frac{7}{12}$
3. 1
4. $\frac{19}{5}$ or $3\frac{4}{5}$
5. 1
6. 8
7. −64
8. 7
9. 19
10. $\frac{15}{2}$ or $7\frac{1}{2}$
11. −61
12. −31
13. 19
14. undefined
15. $-\frac{11}{18}$
16. −39
17. −9
18. −10
19. 0
20. −30
21. $-\frac{9}{23}$
22. 1
23. 20
24. undefined
25. Roody should have first multiplied 5 and 3 instead of adding 2 and 5.
26. −5 should have been multiplied by 4 instead of subtracted from 4.

1.2: Answers

1. -28
2. 16
3. 9
4. -30
5. 5
6. 0.49
7. $6\frac{2}{3}$ or $\frac{20}{3}$
8. 0
9. $1\frac{1}{4}$ or $\frac{5}{4}$
10. -7
11. 1
12. -2.2
13. $8x-32$
14. $-3-18x+12y$
15. $2.4v+2.7$
16. $5-9a$
17. $-3m-n+1$
18. $2n-\frac{1}{2}$
19. $\frac{3}{2}-\frac{10}{3}p+\frac{8}{7}r$
20. $-6a+8b-14$
21. $-8x$
22. $-9b-90$
23. $-\frac{3}{5}x-\frac{2}{5}$
24. $-5x-9$
25. $\frac{3}{8}b$
26. $6p-5$
27. $8p-24$
28. $-10n-9$
29. $-5x+9$
30. $9m-30$
31. $-1.3m+2.4$
32. $-10x+23$
33. 5
34. $6a-4$
35. $5x+32$
36. $-4n+10$
37. $42v$
38. $\frac{14}{9}n+\frac{7}{8}$
39. 0
40. $-28x-19$
41. $-9x+12$
42. $7y-12$

1.3: Answers

1. $v=7$
2. $b=-14$
3. $g=-0.5$
4. $x=\frac{13}{20}$
5. $a=9.7$
6. $k=6\frac{13}{14}$ or $\frac{97}{14}$

Answer Key

7. $h = -19$
8. $p = -6$
9. $n = 12.5$
10. $m = -38$
11. $r = -7$
12. $x = 32$
13. $n = -108$
14. $y = \frac{5}{3}$
15. $v = -80$
16. $x = \frac{1}{3}$
17. $n = \frac{5}{6}$
18. $a = \frac{1}{16}$

19. $x = 10$
20. $k = -78$
21. $n = 3$
22. $x = -21$
23. $y = \frac{5}{4}$
24. $p = -40$
25. $r = \frac{20}{7}$
26. $p = -\frac{1}{40}$
27. $n = -\frac{7}{2}$
28. $c = \frac{5}{12}$
29. Roody added 9 to both sides, but he should have divided both sides by -9.

1.4: Answers

1. $a = 6$
2. $x = 4$
3. $y = -4$
4. $n = 16$
5. $a = \frac{7}{2}$
6. $k = -\frac{3}{2}$
7. $y = 5$
8. $p = 3$
9. $x = \frac{6}{5}$
10. $a = 0$
11. $n = -\frac{9}{2}$
12. $p = -\frac{4}{3}$
13. $y = 4$
14. $x = \frac{1}{7}$

15. $a = 4$
16. $y = -\frac{2}{5}$
17. $n = 0$
18. $k = 4$
19. $x = \frac{19}{5}$
20. $m = 12$
21. $a = -\frac{22}{13}$
22. $y = \frac{4}{3}$
23. $n = -4$
24. $p = \frac{3}{4}$
25. $m = -\frac{13}{3}$
26. $x = -2$
27. $n = 72$
28. $a = -\frac{17}{4}$

1.5: Answers

	Inequality Notation	Number Line Graph	Interval Notation
1.	$n > -5$		$(-5, \infty)$
2.	$x \geqslant 4$ or $4 \leqslant x$		$[4, \infty)$
3.	$x < -2$ or $-2 > x$		$(-\infty, -2)$
4.	$1 \geqslant k$		$(-\infty, 1]$
5.	$x < -5$ or $-5 > x$		$(-\infty, -5)$
6.	$x \leqslant 1$ or $1 \geqslant x$		$(-\infty, 1]$
7.	$-6 \leqslant p$		$[-6, \infty)$
8.	$x > -2$ or $-2 < x$		$(-2, \infty)$
9.	$x \leqslant 5$ or $5 \geqslant x$		$(-\infty, 5]$
10.	$x < 4$		$(-\infty, 4)$

11. $r < 1$; $(-\infty, 1)$

12. $n \leqslant -10$; $(-\infty, -10]$

13. $n \leqslant 5$; $(-\infty, 5]$

14. $n > -6$; $(-6, \infty)$

15. $x \leqslant -18$; $(-\infty, -18]$

16. $m \geqslant 2$; $[2, \infty)$

17. $k > 19$; $(19, \infty)$

18. $r > 8$; $(8, \infty)$

19. $x \leqslant 6$; $(-\infty, 6]$

20. $b > 1$; $(1, \infty)$

21. $n \geqslant 0$; $[0, \infty)$

22. No solution; $\emptyset$

23. All real numbers; $\mathbb{R}$

24. $x > 2$; $(2, \infty)$

25. $v \leqslant 0$; $(-\infty, 0]$

26. Roody has not finished simplifying. His answer $0m \geqslant 7$ can be simplified to $0 \geqslant 7$, which is always false. So the inequality has no solutions.

Answer Key

1.6: Answers

1. $x = 3$ or $x = -3$
2. $y = 0$
3. no solutions
4. $p = 5$ or $p = -5$
5. $n = \frac{38}{9}$ or $n = -6$
6. $c = 15$ or $c = -9$
7. $w = 2$ or $w = -2$
8. $b = \frac{7}{2}$ or $b = -\frac{17}{2}$
9. $x = -6$ or $x = -8$
10. $y = 7$ or $y = -9$
11. $m = 3$ or $m = -\frac{11}{3}$
12. $x = \frac{19}{3}$ or $x = -\frac{19}{3}$
13. no solutions
14. $t = \frac{3}{4}$
15. $g = 5$ or $g = -5$
16. $k = 16$ or $k = 0$
17. $c = \frac{5}{2}$ or $c = -1$
18. no solutions

1.7: Answers

1. $a = P - b - c$
2. $t = \dfrac{I}{pr}$
3. $L = S - 2B$
4. $m = \dfrac{E}{c^2}$
5. $m = \dfrac{P}{n-c}$
6. $w = \dfrac{V}{lh}$
7. $D = \dfrac{12V}{\pi n}$
8. $x = 3 - 5y$
9. $x = \dfrac{1}{a}(c - b)$ or $x = \dfrac{c - b}{a}$
10. $t = \dfrac{1}{a}(c + bw)$ or $t = \dfrac{c + bw}{a}$
11. $h = \dfrac{3V}{\pi r^2}$
12. $a = \dfrac{2A}{h} - b$ or $a = \dfrac{2A - bh}{h}$
13. $a = \dfrac{1}{5}(7b + 4)$ or $a = \dfrac{7b + 4}{5}$
14. $L = \dfrac{1}{6}q + p$ or $L = \dfrac{q + 6p}{6}$
15. $y = -\dfrac{3}{2}x + \dfrac{7}{2}$ or $y = \dfrac{7 - 3x}{2}$
16. $F = \dfrac{9}{5}C + 32$ or $F = \dfrac{9C + 160}{5}$
17. $r = \dfrac{A - p}{pt}$
18. $v = \dfrac{h + 16t^2}{t}$
19. $h = \dfrac{S - \pi r^2}{\pi r}$
20. (a) $y = 180° - x - z$
    (b) $y = 88.2°$
21. (a) $r = \dfrac{C}{2\pi}$
    (b) $r = 5.096$ cm
22. (a) $z = 3A - x - y$
    (b) You must score 81.

23. Roody did not simplify his expression for h. It simplifies to $h = \dfrac{3A}{b}$.

24. On the second step, Roody did not divide the entire right-hand side by a. It should be

$$p = at - b$$
$$\frac{p}{a} = \frac{at - b}{a}$$
$$\frac{p}{a} = \frac{at}{a} - \frac{b}{a}$$
$$\frac{p}{a} = t - \frac{b}{a}$$
$$\frac{p}{a} + \frac{b}{a} = t$$

1.8: Answers

1. Helmet: $81.25; Bike: $650.00
2. The equal sides are 24 inches each and the third side is 12 inches.
3. $226.60
4. 7 or fewer water bottles
5. 3 feet and 5 feet
6. at least 91 cups
7. 430 cubic feet
8. 15 boys and 30 girls
9. 960 bags
10. 30°, 60°, 90°
11. 49 meters and 27 meters
12. at least 23 cups
13. 25,200
14. $4000
15. 93
16. Jack: $2500; Jim: $5000
17. $8\frac{1}{2}$ hours
18. 9 years
19. $525,000

20. 10 lawns

21. (a) Yes, it really works for any number x.

    (b) If you choose a number x, the puzzle gives the expression $\frac{2x+6}{2} - x$. But this simplifies to 3, no matter what x is.

22. Many other puzzles are possible.

2.1: Answers

1. A (3,5); B (0,3); C (−4,1); D (4,0); E (−2,−2); F (0,−6)

2. P (−10,10); Q (5, frac52); R (0,−1); S (9,0); T (0, −$\frac{15}{2}$); U (−10,−10)

3. Quadrant I
4. Quadrant II
5. Quadrant IV
6. y-axis
7. Quadrant III
8. y-axis
9. x-axis
10. Quadrant I
11. Quadrant III
12. x-axis

2.2: Answers

1.

x	y
−1	−2
0	0
1	2

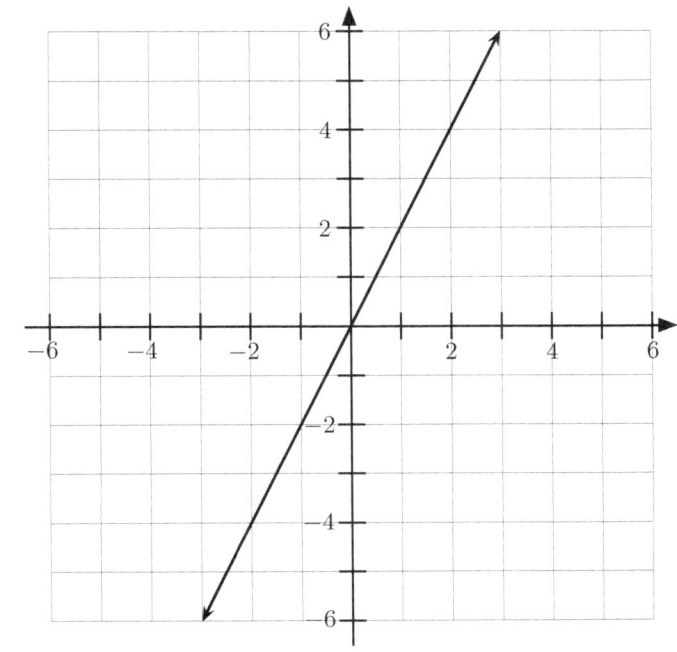

2.

x	y
−3	5
−1	3
2	0

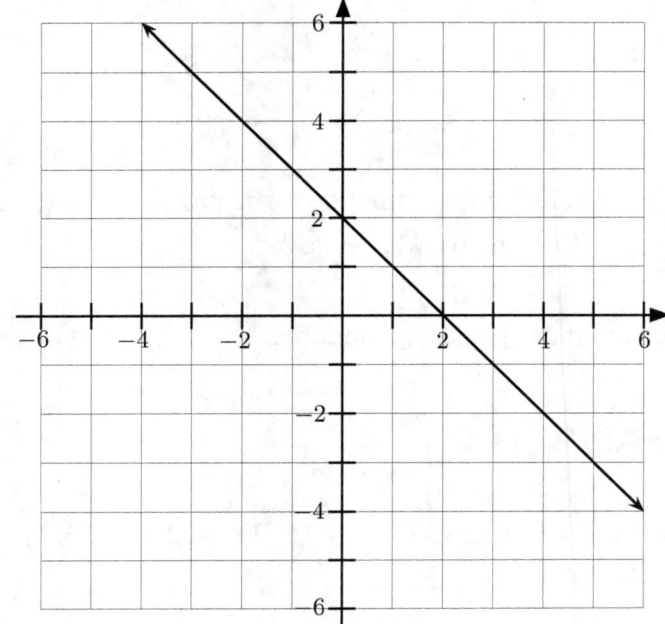

Answer Key 339

3.

x	y
−2	1
1	2
4	3

4.

x	y
−2	4
−1	4
3	4

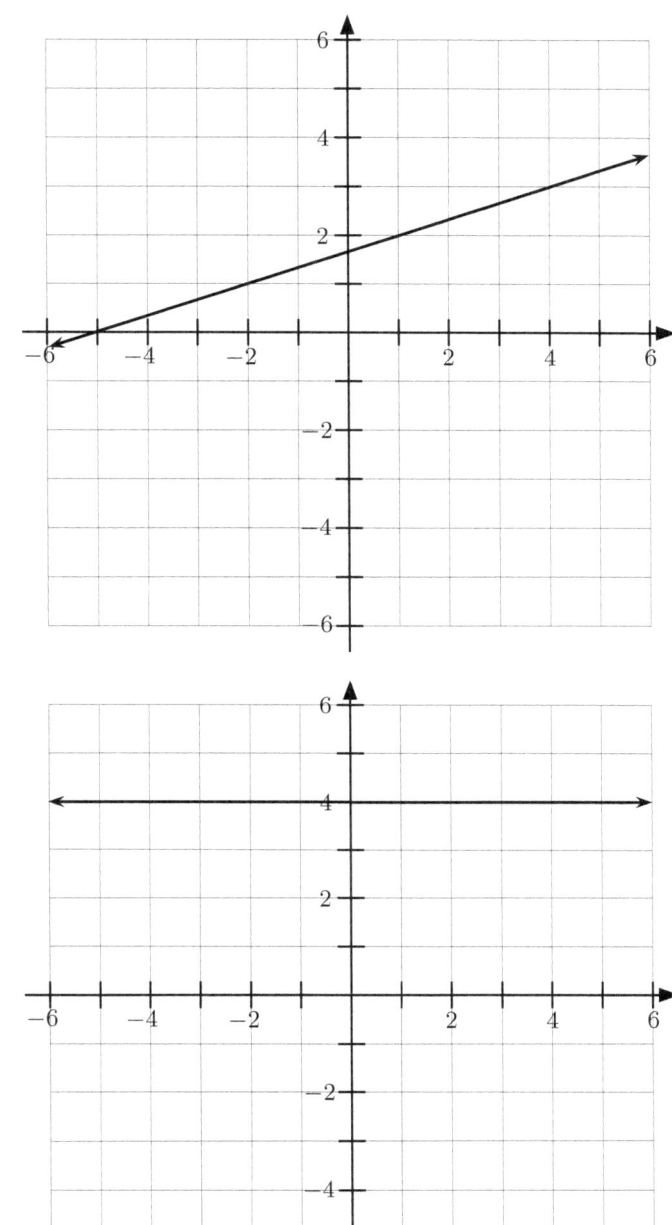

5.

x	y
−3	−4
−3	1
−3	2

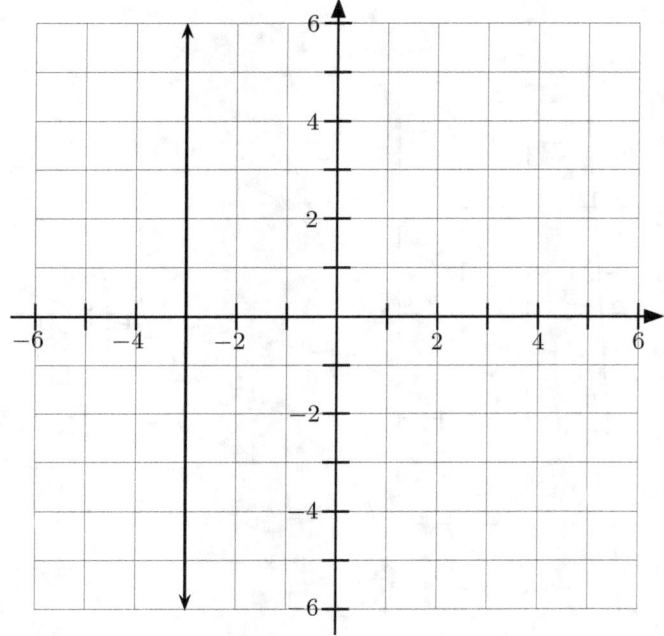

6. There are many ways to fill complete the table. For example,

x	y
−2	5
0	1
2	−3

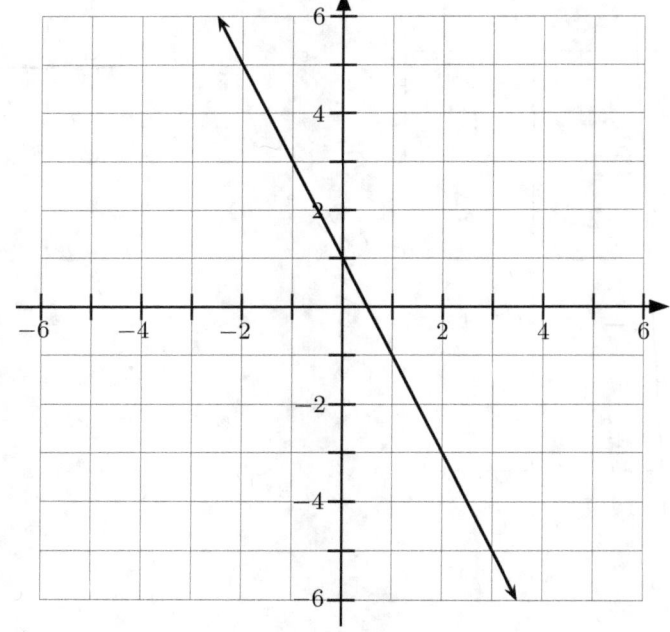

2.3: Answers

1. x-intercept: $(0,0)$ y-intercept: $(0,0)$

2. x-intercept: $(4,0)$
   y-intercept: $(0,4)$

3. x-intercept: does not exist
   y-intercept: $(0,2)$

4. x-intercept: $(-2,0)$
   y-intercept: does not exist

Answer Key

5. x-intercept: (4, 0)
   y-intercept: (0, −3)

6. x-intercept: (−1, 0)
   y-intercept: (0, 3)

7.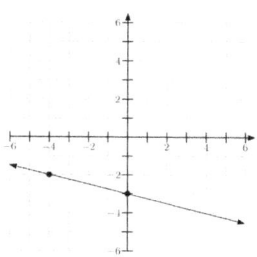

11.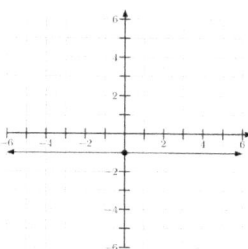

8.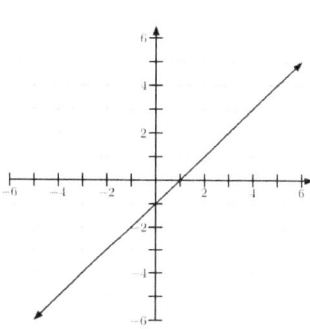

12.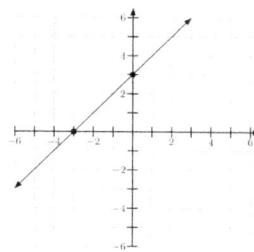

13.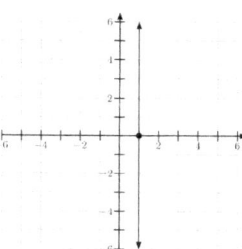

9.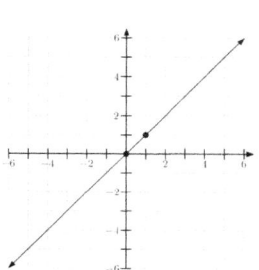

14.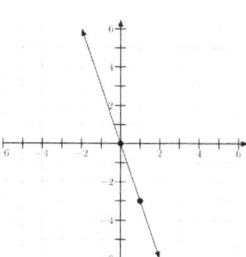

10.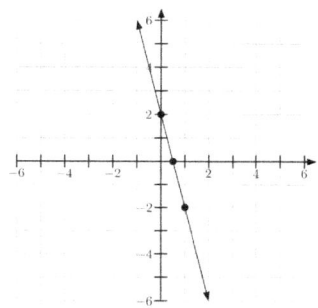

15.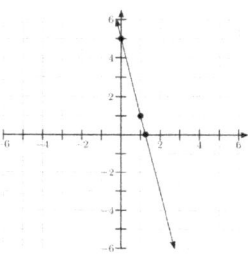

16.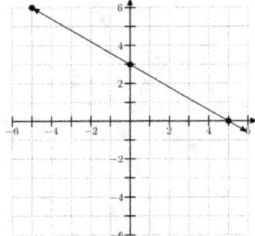

17.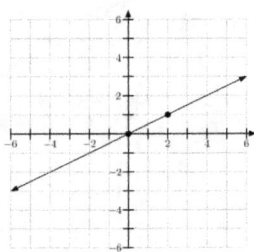

18.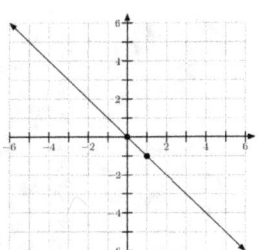

19.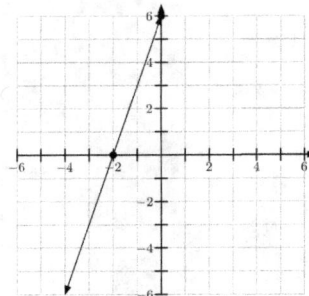

20.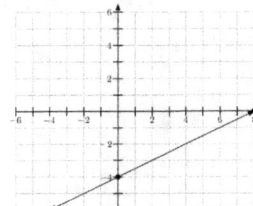

2.4: Answers

1. $m = \dfrac{3}{2}$

2. Undefined Slope

3. $m = \dfrac{5}{6}$

4. $m = -1$

5. $m = \dfrac{10}{3}$

6. $m = -\dfrac{1}{2}$

7. $m = -\dfrac{2}{3}$

8. $m = \dfrac{5}{4}$

9. $m = -1$

10. $m = 0$

Answer Key

11. Undefined Slope

12. $m = \dfrac{16}{7}$

13. $m = -\dfrac{17}{31}$

14. $m = -\dfrac{3}{2}$

15. $m = \dfrac{4}{3}$

16. $m = -\dfrac{7}{17}$

17. $m = 0$

18. $m = \dfrac{5}{11}$

19. $m = \dfrac{1}{2}$

20. $\dfrac{1}{16}$

21. $y = -5$

22. $y = 2$

23. $y = 3$

24. $y = 1$

25. Many solutions, example: $(8, 15)$

26. Many solutions, example: $(3, -2)$

27. Many solutions, example: $(-2, 4)$

28. Many solutions, example: $(1, 2)$

29. Many solutions, example: $(-3, 12)$

30. Many solutions, example: $(3, -3)$

31. (a) 0.5 mL/lb

    (b) 17.5 mL

32. $34,000 per year

33. 193007.33 tons of recycled or diverted waste per year

34. Roody did not follow the same order for the numerator and denominator.

35. Yes

36. No

2.5: Answers

1. $y = 2x + 5$
2. $y = -6x + 4$
3. $y = x - 4$
4. $y = -x - 2$
5. $y = -\frac{3}{4}x - 1$
6. $y = -\frac{1}{4}x + 3$
7. $y = \frac{1}{3}x + 1$
8. $y = \frac{2}{5}x + 5$
9. $y = -x + 5$
10. $y = -3x - 5$
11. $y = x - 1$
12. $y = -\frac{5}{3}x - 3$
13. $y = -4x$
14. $y = -\frac{3}{4}x + 2$
15. $y = -\frac{x}{10} - \frac{37}{10}$
16. $y = \frac{x}{10} - \frac{3}{10}$
17. $y = -2x - 1$
18. $y = \frac{6}{11}x + \frac{70}{11}$
19. $y = \frac{7}{3}x - 8$
20. $y = -\frac{4}{7}x + 4$
21. Does not exist
22. $y = \frac{1}{7}x + 6$
23. $y = -x - 1$
24. $y = \frac{5}{2}x$
25. $y = 4x$
26. $y = -\frac{2}{3}x + 1$
27. $y = -4x + 3$
28. Does not exist
29. $y = -\frac{1}{2}x + 1$
30. $y = \frac{6}{5}x + 4$

Answer Key

31.
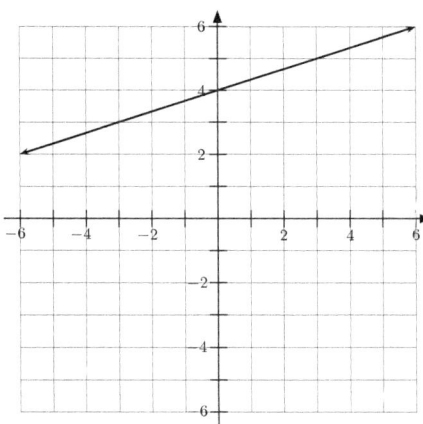

32.
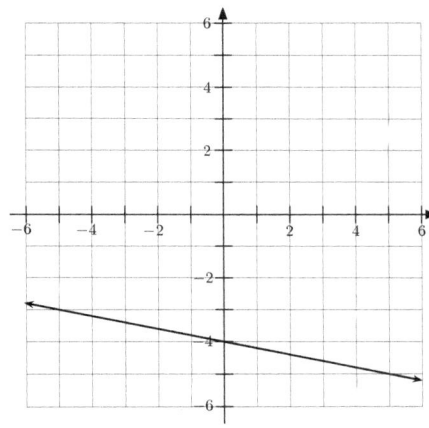

33.
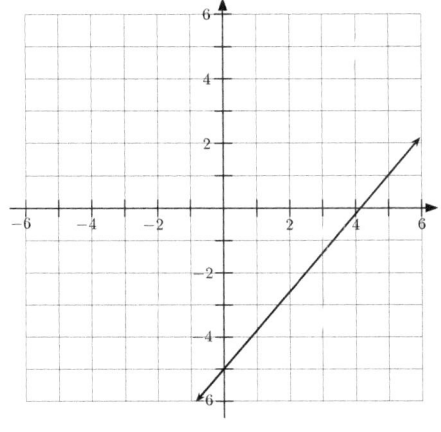

34.
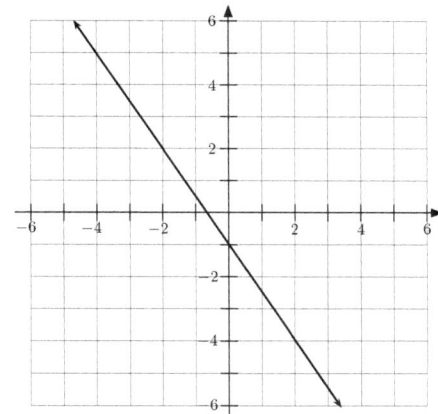

35.
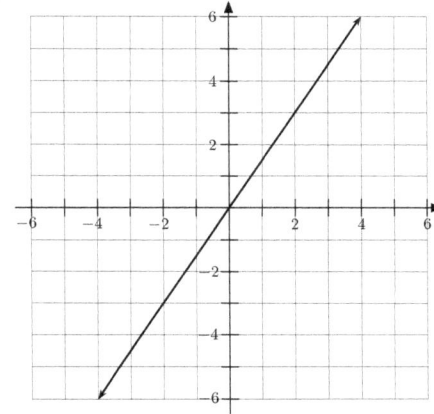

36.
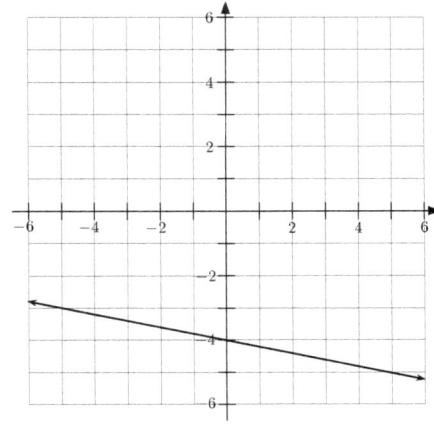

37.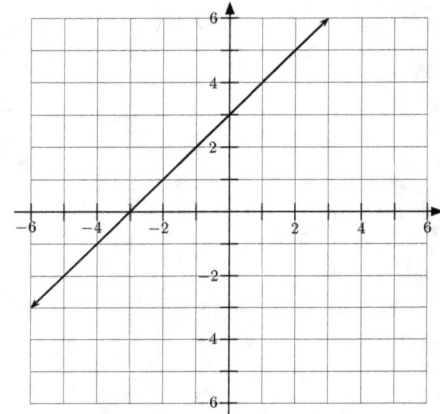

38.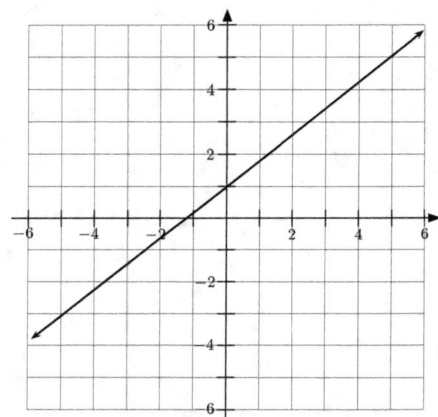

39.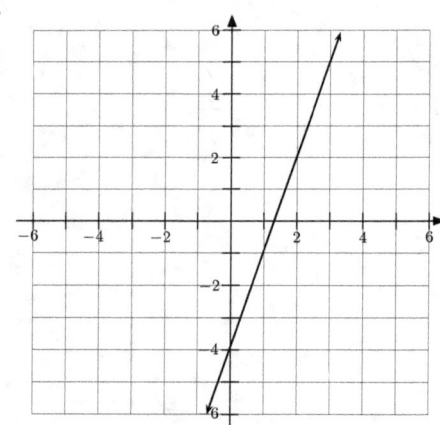

40.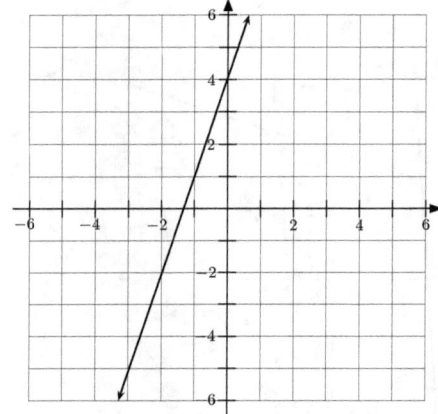

41.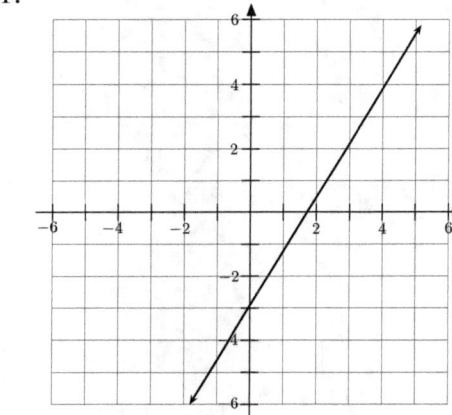

42.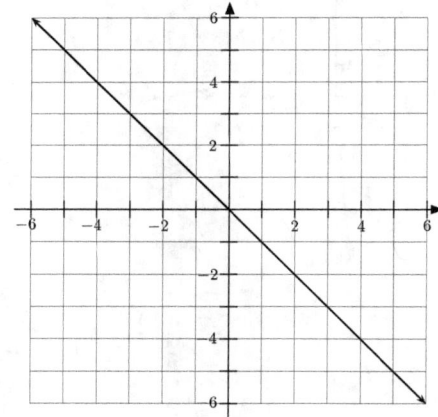

43. (a) $y = -2x + 250$
    (b) 245 lbs
    (c) 37.5 weeks

44. (a) $y = -\dfrac{3}{2}x + 30$

(b) $(0, 30)$

(c) When Leigh starts the program, he smokes 30 cigarettes per day.

(d) 20 weeks

2.6: Answers

1. $y+5 = 9(x+1)$
2. $y-2 = \dfrac{1}{2}(x-2)$
3. $y+7 = -\dfrac{1}{4}x$
4. $y = 1$
5. $x = 2$
6. $y = -\dfrac{5}{4}(x+1)$
7. $y-3 = -2(x+4)$
8. $y = 6$
9. $y+1 = -(x+4)$
10. $x = -8$
11. $y+2 = \dfrac{3}{2}(x+4)$
12. $y-5 = \dfrac{1}{4}(x-3)$
13. $y = x-4$
14. $y = 2x-5$
15. $x = 3$
16. $y = -\dfrac{3}{2}x+6$
17. $y = -\dfrac{3}{4}$
18. $y = -\dfrac{2}{3}x - \dfrac{10}{3}$
19. $y = -x+2$
20. $y = \dfrac{1}{3}x+1$
21. $y = -x+5$
22. $x = \dfrac{2}{7}$
23. $y = -\dfrac{3}{4}x - \dfrac{11}{4}$
24. $y = \dfrac{1}{2}x - \dfrac{3}{2}$
25. $y = -\dfrac{1}{2}x + \dfrac{3}{2}$
26. $x = 4$
27. $y = -2$
28. $y = \dfrac{2}{3}x - \dfrac{4}{3}$
29. (a) $(1, 68000), (4, 272000)$
    (b) $m = 68000$. The company's profit is increasing at a rate of \$68,000 per year.
    (c) $y = 68000x$
    (d) \$408,000
30. (a) $(2, 1197.68), (7, 4191.88)$
    (b) $m = 598.84$. Tuition is incrasing \$598.84 per credit for international students.
    (c) $y = 598.84x$
    (d) \$5988.40
31. (a) $(1, 17530), (3, 11849)$
    (b) $m = -2840.5$
    (c) Every year, the Kia Soul's value decreases by \$2840.50.
    (d) $y = -2840.50x + 20370.5$
    (e) $(0, 20370.5)$
    (f) The original value of the car was \$20,370.50.

2.7: Answers

	Given Line	Slope of Parallel Line	Slope of Perpendicular Line
1.	$y = 4x - 5$	4	$-\frac{1}{4}$
2.	$x = y + 4$	1	-1
3.	$3y = x + 6$	$\frac{1}{3}$	-3
4.	$x + 2y = 8$	$-\frac{1}{2}$	2

5. $y = \frac{9}{2}x - \frac{19}{2}$

6. $y = -\frac{3}{4}x - \frac{1}{4}$

7. $y = 2x + 1$

8. $y = -\frac{5}{2}x + 2$

9. $y = -x - 4$

10. $y = \frac{7}{5}x + \frac{13}{5}$

11. $y = -3x$

12. $y = -\frac{1}{2}x - 3$

13. $x = 2$

14. $y = -2x$

15. $y = \frac{1}{5}x + 1$

16. $x = -8$

17. $y = x + 2$

18. $y = -3$

19. $y = -\frac{1}{2}x + 4$

20. $y = \frac{7}{3}x + 2$

21. Roody cannot use the point slope form. He can use the fact that a vertical line is perpendicular to a horizontal line, so he is looking for a vertical line that goes through the point $(-5, 3)$, which is the line $x = -5$.

2.8: Answers

1.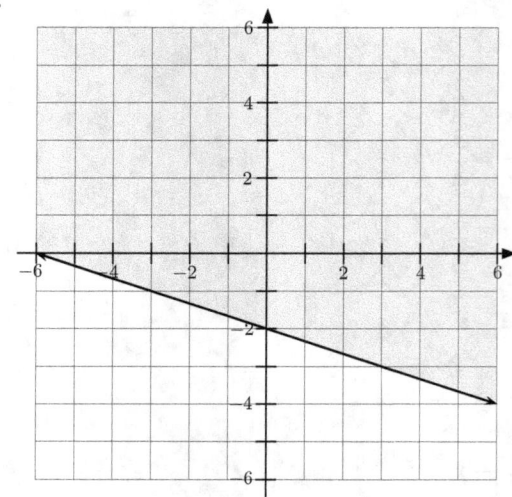

Answer Key

2.
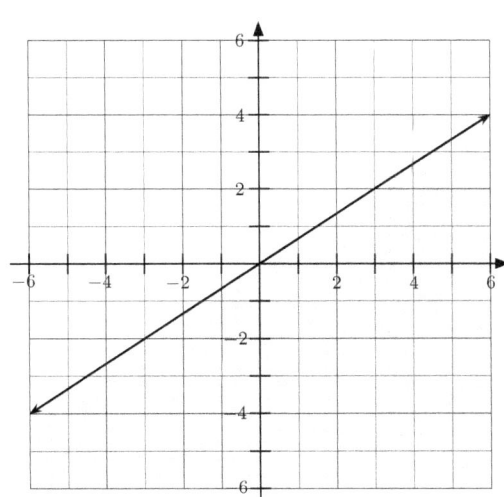

3.
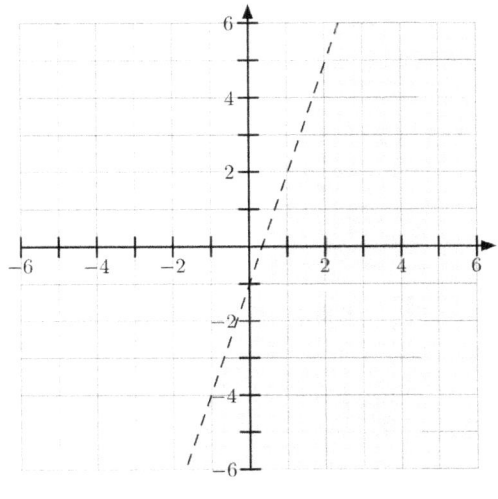

4.
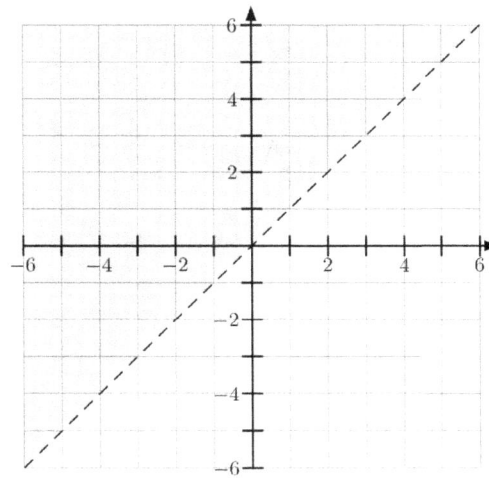

5.
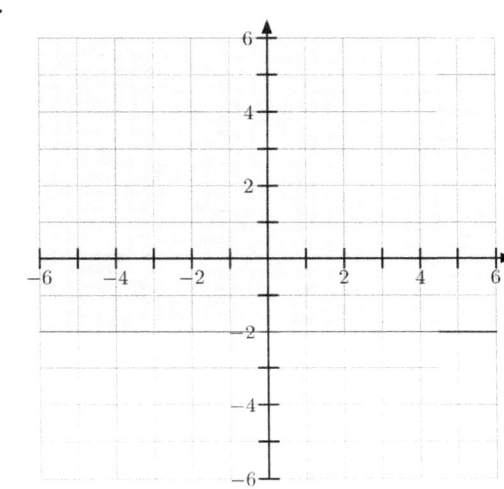

6.
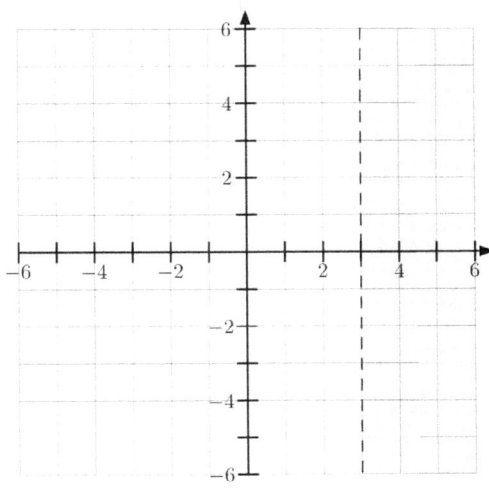

350   Answer Key

7.
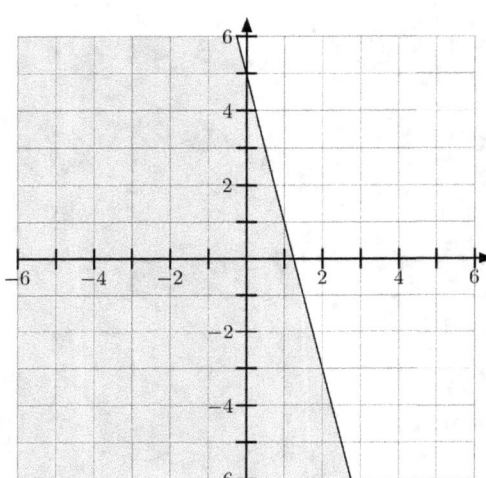

10.
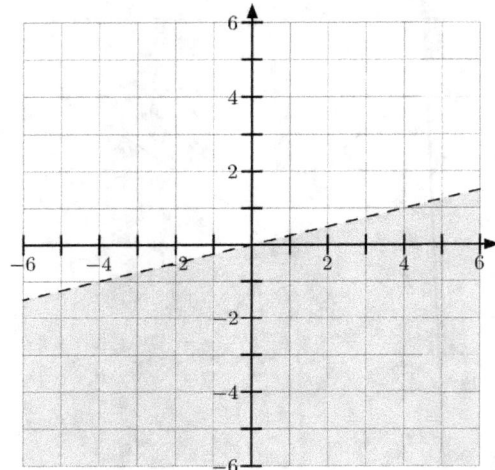

8.
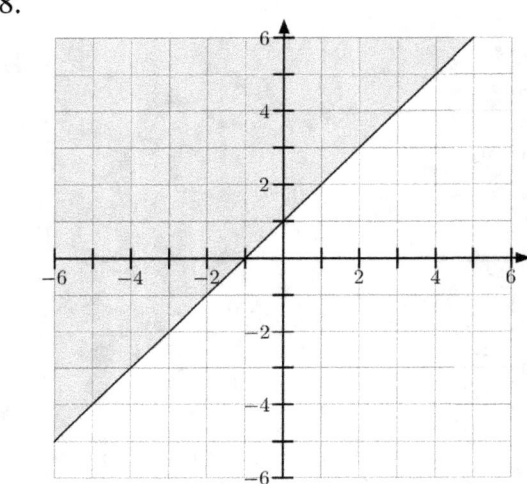

11.
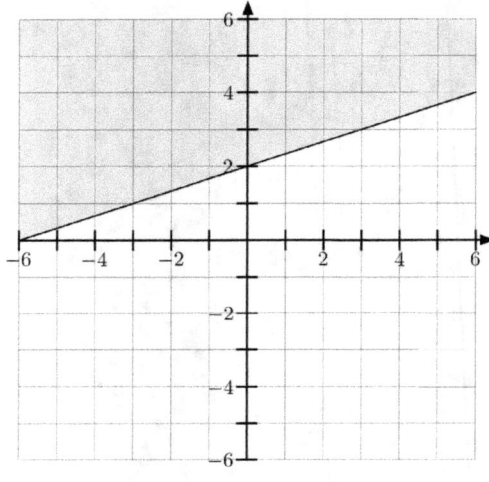

9.
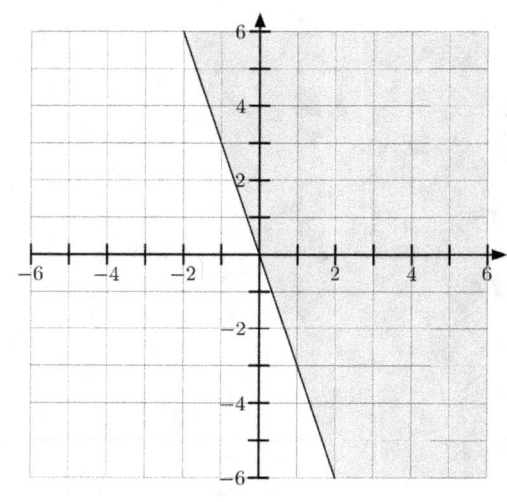

12.
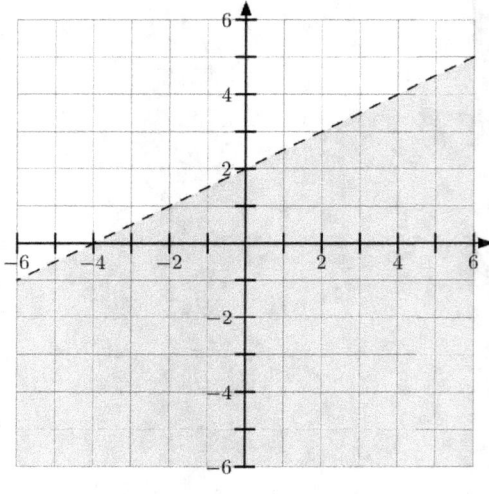

Answer Key   351

13. 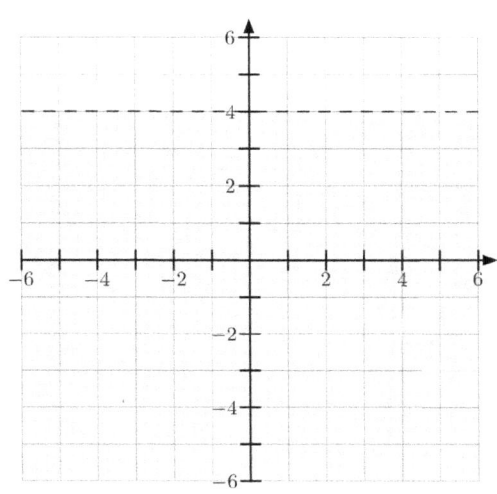   14.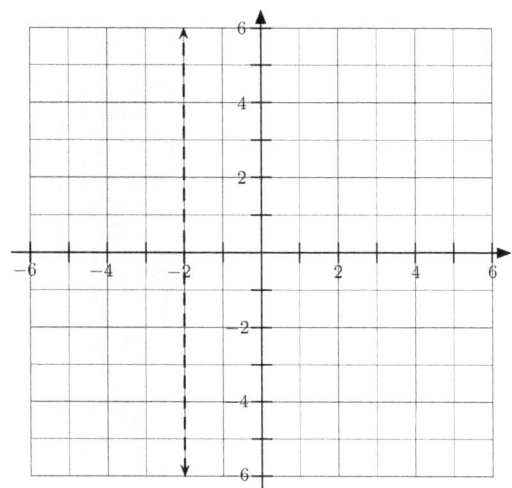

3.1: Answers

1. Yes   6. Yes

2. Yes   7. No

3. No   8. No

4. Yes   9. Yes

5. Yes   10. No

3.2: Answers

1. $(1, 2)$ 2. $(-2, 3)$

3. $(-4, 1)$

4. No solution

5. $(3, -1)$

6. Infinitely many solutions

7. $(2, 2)$

8. $(3, -2)$

9. $(4, 0)$

10. $(1, -2)$

3.3: Answers

1. $(x, y) = (1, -3)$
2. $(a, b) = (-3, 2)$
3. $(p, q) = (0, 3)$
4. $(g, h) = (-45, -11)$
5. $(m, n) = (0, 2)$
6. $(x, y) = (-1, 8)$
7. $(m, n) = (3, 7)$
8. $(c, d) = \left(\frac{1}{3}, \frac{49}{3}\right)$
9. $(p, q) = (8, -8)$
10. $\varnothing$ or No solution
11. $(a, b) = (-3, -2)$
12. $(c, d) = (-6, 17)$
13. $(x, y) = (2, 1)$
14. $\varnothing$ or No solution
15. $(x, y) = (-4, 3)$
16. $(a, b) = \left(\frac{3}{2}, 2\right)$
17. Infinitely many solutions
18. $(x, y) = (0, 3)$
19. $(m, n) = (1, -4)$
20. $(x, y) = \left(\frac{11}{2}, -\frac{3}{2}\right)$
21. $(m, n) = (2, 0)$
22. $(a, b) = (-4, 8)$

Answer Key

23. (a) $(2,0)$
    (b) $y = -x + 2$ and
        $y = \frac{3}{2}x - 3$
    (c) $(2,0)$ is the correct solution.

24. (a) No solutions
    (b) $y = -\frac{3}{2}x + 3$ and $y = -\frac{3}{2}x - 3$
    (c) Substitution finds no solutions.

25. (a) $(1,3)$
    (b) $y = 3$ and $-3x + 2y = 3$
    (c) $(1,3)$ is the correct solution.

26. (a) $(3,-2)$
    (b) $x = 3$ and $y = -\frac{2}{3}x$
    (c) $(3,-2)$ is the correct solution.

3.4: Answers

1. $(x,y) = (2,4)$
2. $(m,n) = (2,-4)$
3. $\varnothing$ or No solution
4. $(c,d) = (-2,-9)$
5. $(u,v) = (1,-2)$
6. $(x,y) = (-1,0)$
7. $\varnothing$ or No solution
8. $(x,y) = (-6,-8)$
9. Infinitely many solutions
10. $(a,b) = (2,-2)$
11. $(c,d) = (1,2)$
12. $(g,h) = (2,\frac{1}{2})$
13. $(a,b) = (-\frac{1}{2}, \frac{2}{3})$
14. Infinitely many solutions
15. $(x,y) = (0,4)$
16. $(x,y) = (-\frac{2}{3}, \frac{3}{4})$
17. $(x,y) = (-3,6)$
18. $(x,y) = (-20,3)$
19. $(x,y) = (-2,-3)$
20. $(x,y) = (-\frac{1}{2}, -1)$

3.5: Answers

1. Mary weighs 34 lbs and Bobby weighs 41 lbs.
2. 23 snow shovels and 115 flashlights.
3. prep time 25 minutes and bake time is 40 minutes
4. 1.8 gallons of pure alcohol
5. 16.8 ounces of 26¢ per ounce tea; new mixture will weigh 28.8 ounces
6. 10 kilograms of root beer barrels; new mixture will weight 34 kilograms
7. 36 lbs. of dog food that is 8% fat and 12 lbs. that is 20% fat
8. 110 quarts of pure maple syrup; 40 quarts of 85% maple syrup

9. An adult ticket is $22.55; a youth ticket is $16.55

10. 18 ounces of peanuts and raisins mix; 10 ounces of chocolate candies

11. Cheese is $3.25 a pound; Ham is $6.50 a pound

12. 83 adults, 129 children

13. 28 burgers and 14 veggie dogs

14. Freight train traveled at 30 mph; Passenger train traveled at 45 mph

15. Jogs $\frac{3}{4}$ hour (or 45 minutes); Walk $\frac{1}{4}$ hour (or 15 minutes)

16. (a) 4 hours going to airport; 3 hours returning from airport;
    (b) 360 miles

17. Mark's jogging speed is 2.2 mph and Jackies' rollerblading speed is 6.2 mph

18. $\frac{1}{3}$ hour or 20 minutes

4.1: Answers

1. $w \cdot w \cdot w \cdot w$
2. $(-w)(-w)$
3. $-w \cdot w$
4. $(2w)(2w)(2w)(2w)$
5. $2 \cdot w \cdot w \cdot w \cdot w$
6. $\left(\frac{5}{p}\right)\left(\frac{5}{p}\right)\left(\frac{5}{p}\right)$
7. $\frac{5 \cdot 5 \cdot 5}{p}$
8. $\frac{5 \cdot 5 \cdot 5}{p \cdot p \cdot p}$
9. $\frac{5}{p \cdot p \cdot p}$
10. $(x-7)(x-7)$
11. $x \cdot x - 7 \cdot 7$
12. $g \cdot g \cdot g$
13. $-g \cdot g \cdot g$
14. $(-g)(-g)(-g)$
15. $7 \cdot g \cdot g \cdot g \cdot g$
16. $(7g)(7g)(7g)(7g)(7g)$
17. $\frac{g}{-7 \cdot 7 \cdot 7 \cdot 7}$
18. $\frac{g \cdot g \cdot g \cdot g \cdot g}{7}$
19. $\left(\frac{g}{-7}\right)\left(\frac{g}{-7}\right)\left(\frac{g}{-7}\right) \cdot \left(\frac{g}{-7}\right)\left(\frac{g}{-7}\right)$
20. $(k+2)(k+2)(k+2)(k+2)$
21. $k \cdot k \cdot k \cdot k + 2 \cdot 2 \cdot 2 \cdot 2$

4.2: Answers

1. 9^7
2. x^6
3. $12x^3$
4. $12m^2n$
5. $8m^6n^3$
6. x^3y^6
7. 7^8
8. y^{12}
9. x^3y^3
10. $16a^{16}$
11. $16x^4y^4$
12. $4u^6v^4$
13. p^2
14. 3
15. $2y^3$

Answer Key

16. $\frac{m^2}{2}$
17. $2xy^3$
18. $\frac{4x^2y}{3}$
19. $\frac{y^2}{2}$

20. 9
21. $49m^{22}$
22. $2x^{17}y^{16}$
23. 64
24. $\frac{2}{a^6}$

25. $\frac{x^2y^5}{2}$
26. $\frac{m^2n^7}{2}$
27. x^3y^3
28. $16x^4y^{12}$

4.3: Answers

1. 2
2. 1
3. 25
4. −25
5. 25
6. $\frac{1}{25}$
7. 64
8. $\frac{1}{64}$
9. −64
10. −64
11. $\frac{7}{8}$
12. $1\frac{1}{9}$ or $\frac{10}{9}$
13. $\frac{5}{6}$
14. $1\frac{3}{5}$ or $\frac{8}{5}$
15. $-\frac{1}{9}$
16. $\frac{1}{9}$

17. $-\frac{1}{x^4}$
18. $\frac{1}{x^4}$
19. $\frac{3}{w^5}$
20. x^{20}
21. r
22. $\frac{1}{y^5}$
23. 4
24. $\frac{k^4}{5}$
25. $8b^6$
26. $\frac{4}{5}$
27. 1
28. $\frac{-8}{n^{18}}$
29. $\frac{16}{y^4}$
30. $\frac{-4}{c^{11}}$
31. $6x^4$

32. $\frac{y^{12}}{4}$
33. $-\frac{1}{2}$
34. $-\frac{1}{2a^{16}}$
35. >
36. <
37. <
38. =
39. =
40. =
41. >
42. >
43. >
44. =
45. 1/4
46. 1/8
47. $\frac{-6}{x^3}$

4.4: Answers

1. 8.85×10^3
2. 8.1×10^{-2}
3. 3.91×10^{-6}
4. 7.44×10^{-4}
5. 1.09×10^6
6. 1.5×10^{10}

7. 870000

8. 0.0009

9. 200,000,000

10. 256

11. 53,300

12. 0.000067

	Name	Standard Notation	Scientific Notation
13.	Trillion	1,000,000,000,000	1×10^{12}
14.	Billion	1,000,000,000	1×10^{9}
15.	Million	1,000,000	1×10^{6}
16.	Thousand	1,000	1×10^{3}
17.	Tenth	0.1	1×10^{-1}
18.	Hundredth	0.01	1×10^{-2}
19.	Millionth	0.000001	1×10^{-6}
20.	Billionth	0.000000001	1×10^{-9}

21. <

22. >

23. >

24. >

25. <

26. >

27. 9.4×10^{2}

28. 9.61×10^{-12}

29. 3.45×10^{-5}

30. 2.4×10^{3}

31. $\$1.9 \times 10^{13}$

32. 8.8×10^{19} bytes

33. 1.25×10^{-2} seconds/beat

34. $\$4.5 \times 10^{8}$ or $450,000,000

35. 1.0×10^{12}

36. 1.16×10^{4} or 11,600

37. Roody read the number backwards. It should be 3.25×10^{-3}

38. Roody moved the decimal point the wrong way. It should be 0.0000123.

5.1: Answers

Answer Key

1. Polynomial
2. Not a Polynomial
3. Polynomial
4. Not a Polynomial
5. Polynomial
6. Not a Polynomial

	Expression	Number of Terms	Degree	Type of Polynomial
7.	$x^2 + 9.1x + 2.3$	3	2	Trinomial
8.	$3n - 4n^4$	2	4	Binomial
9.	$-\dfrac{5}{8}$	1	0	Monomial
10.	$n^3 - 7n^2 + 15n - 20$	4	3	Polynomial
11.	$2p$	1	1	Monomial
12.	$n^7 + n^3$	2	7	Binomial

5.2: Answers

1. $4x^2$
2. $2x^2 - x$
3. $6x^3 + 4x^2$
4. $3k^2 + 9k$
5. $8x^3 + 8x^2$
6. $-3x$
7. $6n^4 - 5n^2 - 1$
8. $p^2 + 4p - 6$
9. $5b^3 + 12b^2 - 3b + 8$
10. $3m^2 - 2m + 6$
11. $10x^3 + 3x - 8$
12. $9n^4 + 4n^3 + 4$
13. $4b^3 - b^2 - 11b + 19$
14. $14n^3 + 3n^2 - 6n + 8$
15. 0
16. $7x^3 + 9x^2 + 4x$
17. x^2 and x are not like terms and connot be combined. The correct answer is $x^2 - x$.
18. Roody did not distribute the "$-$" in the middle to the 7. The correct answer is $2x - 12$.

5.3: Answers

1. $6p-42$
2. $12x^2+6x$
3. $20m^5+20m^4$
4. $c^2-2c-15$
5. $x^2+8x+15$
6. $15v^2-26v+8$
7. $24x^2-22x-7$
8. $3x^2+13xy+12y^2$
9. a^2-b^2
10. $9-y^2$
11. $25n^2-16$
12. $9p^2-42p+49$
13. $a^2+2ab+b^2$
14. $w^3+6w^2+12w+8$
15. c^3-3c^2+3c-1
16. $a^3-3a^2b+3ab^2-b^3$
17. $6r^3-43r^2+12r-35$
18. $7x^2-49x+70$
19. $96x^2-6$
20. $12n^3-20n^2+38n-20$
21. $4x^2-18x+9$
22. $36-20y-9y^2$
23. $6x+37$
24. $3x^2-6x$
25. Roody did not multiply out $(x-9)(x-9)$. He forgot the x terms. The correct answer is $x^2-18x+81$.
26. Roody forget the x terms when multiplying. The correct answer is $x^2+16x+64$.

5.4: Answers

1. m^2+2m+3
2. $3x^3+8x^2-2x-1$
3. $5x+\frac{1}{4}+\frac{1}{2x}$
4. $2n^2+5n+1+\frac{4}{n}$
5. $\frac{3k^2}{8}+\frac{k}{2}+\frac{1}{4}+\frac{1}{k}$
6. $\frac{5}{9}x^2+5x^2+\frac{4}{9}x+1$
7. $a+4+\frac{2}{a-8}$
8. $v+8$
9. $n+3-\frac{7}{n+4}$
10. $2x+3$
11. $2r-1+\frac{10}{2r+5}$
12. x^2-8x+7
13. $k^2-3k-9-\frac{5}{k-1}$
14. $n+2$
15. $3v+1+\frac{3}{3v-1}$
16. $a^2-2a-11$
17. $4m^2-m-3-\frac{18}{m-3}$
18. n^2+4n-6

www.ingramcontent.com/pod-product-compliance
Lightning Source LLC
Chambersburg PA
CBHW062318220526
45469CB00008B/2547